T0202712

Guideline for EN 9100:2018

Martin Hinsch

Guideline for EN 9100:2018

An Introduction to the European
Aerospace and Defence Standard

 Springer

Martin Hinsch
Hamburg, Germany

ISBN 978-3-662-61366-5   ISBN 978-3-662-61367-2   (eBook)
https://doi.org/10.1007/978-3-662-61367-2

Planung/Lektorat: Michael Kottusch
This Springer imprint is published by the registered company Springer-Verlag GmbH, DE part of
Springer Nature.
The registered company address is: Heidelberger Platz 3, 14197 Berlin, Germany

Preface

Companies that design, produce or maintain aeronautical products are subject to national aviation legislation all over the world, while being simultaneously monitored by the responsible aviation supervisory authorities. However, this only applies to some extent to the company's suppliers. In order to create an appropriate and comparable quality level for these companies as well, EN 9100 was published in 2003 as a certifiable system standard. Since then, it has rapidly spread within the aviation industry and is now considered as an obligatory foundation of operational quality management for nearly every market participant.

This book is intended to raise a fundamental awareness of the requirements of EN 9100:2018. At the same time, this text supplements the little literature in the field of aviation QM systems and complements the sparse literature in the field of aeronautical QM systems. Thus, this book can help to develop an appropriate understanding of the structure of companies in the aerospace industry. Furthermore, the following chapters are also useful as a guide for those companies that are seeking a regulatory approval according to EASA Part 21 or 145.

As a precaution, I would like to point out to the reader that QM system standards offer a lot of room for interpretation. If suitable, tips for the implementation are given in the following chapters, they are derived from experience in conformity with the standard, which I have gained in numerous EN 9100 projects as a consultant or auditor. In this respect, the book aims to focus on consistent practical orientation. However, this wide scope for implementation also means that the perception and assessment of a certification auditor can occasionally deviate from the tips and information provided by this book. Thus, there will be EN auditors who interpret the standard or individual sections more strictly, but also those who interpret EN 9100 less strictly.

Unfortunately, the wording in almost all standards is artificial and very formal, though not always immediately accessible to a layperson. This text is intended to assist with the translation of the standard to the language of everyday business. I therefore hope, to have presented the text in such a way that it not only benefits QM experts, but that it is also understandable for practitioners and students without previous QM knowledge. For reasons of simplicity, the text is structured in the same way as EN 9100 from chap. 4 onwards. Whenever it seemed sensible, this has been broken down and reviewed according of the standard. For copyright reasons, it was not possible to reproduce the original text of the standard. In this respect, this book is only an additive, but not an alternative to the actual EN 9100 text of the standard.

I would like to thank all of you who helped me during the six months of writing this book. Especially, I would like to thank Dr. Dorit Sobotta for her expert advice on translating the manuscript.

For further information or support on EN 9100 or its certification, please visit www.aeroimpulse.de or contact me at mh@aeroimpulse.de.

Hamburg   Prof. Dr. Martin Hinsch
Spring 2020

Contents

Abbreviations

4F	Form, Fit, Function, Fatigue
AD	Airworthiness Directive
AECMA	European Association of Aerospace Industries
AMC	Acceptable Means of Compliance
AMM	Aircraft Maintenance Manual
ATP	Acceptance Test Procedure
Chap.	Chapter
CM	Configuration Management
CMM	Component Maintenance Manual
CofC	Certificate of Conformity
CRM	Customer Relationship Management
CS	(EASA) Certification Specification
EAQG	European Aerospace Quality Group
EASA	European Aviation Safety Agency
EN	European Norm
ESD	Electrostatic Discharge
FAA	Federal Aviation Administration
FAI	First Article Inspection
FAIR	First Article Inspection Report
FAR	(US) Federal Aviation Regulations
FMEA	Failure Mode and Effect Analysis
FOD	Foreign Object Debris
GTC	General Terms and Conditions
IAQG	International Aerospace Quality Group
IP	Intellectual Property
IPC	Illustrated Parts Catalogue
ISO	International Organisation for Standardisation
KPI	Key Performance Indicator
MRB	Material Review Board
NAA	National Aviation Authority
NCR	Nonconformity Report
NDT	Non-Destructive Testing
OHSAS	Occupational Health- and Safety Assessment Series
PDCA	Plan-do-Check-Act

PEAR	Process Effectiveness Assessment Report
PO	Purchase Order
OASIS	Online Aerospace Supplier Information System (of the IAQG)
OEM	Original Equipment Manufacturer
QM	Quality Management
QMR	Quality Management Representative
QMM	Quality Management Manual
QMS	Quality Management System
OTD	On-Time-Delivery
OTQ	On-Target Quality
RM	Risk Management
RTCA	Radio Technical Commission for Aeronautics
SB	Service Bulletin
SME	Small and Medium Sized Enterprises
SRM	Structural Repair Manual
SUP	Suspect Unapproved Parts

Introduction to Standardisation and the QM System According to EN 9100

1

1.1 Fundamentals and History of ISO Management Systems

Standardisation is a systematically initiated common harmonisation of processes, systems, terms or product characteristics for the benefit of a user group. The creation of standards is a uniform approach which, on one hand, introduces quality measures and thereby comparability. On the other hand, standardisation increases efficiency by eliminating uncertainties in planning as well as reducing technical and financial adjustments, thus simplifying the movement of goods and services.[1] For this purpose, the following types of standardisation are distinguished:

(a) procedural standards (e.g. quality management according to ISO 9000),
(b) technical standards (e.g. screw type, DIN A4) and
(c) classification standards (e.g. country codes such as de, com, jp).

In order to develop their effectiveness, norms do not have to be formally and legally binding. The fact that most market participants follow a standard, also disciplines those who initially did not meet its requirements. Many standards exert (voluntary) pressure and thus have a stronger effect than laws: Whoever does not obey them will be punished by the market.

The first international standardisation efforts were already undertaken at the end of the nineteenth century/beginning of the twentieth century and had increased thereon rapidly. High growth took place especially after the Second World War with the foundation of the International Organisation for Standardisation (ISO), a sub-organisation of the UNO.

---

[1]See Hinsch (2019) p. 36.

© Springer-Verlag GmbH Germany, part of Springer Nature 2020
M. Hinsch, *Guideline for EN 9100:2018*,
https://doi.org/10.1007/978-3-662-61367-2_1

Until the nineteen seventies, the development and dissemination of technical standards dominated. It was not until 1979 that a standard for quality management systems was published for the first time. The ISO 9000 series of standards emerged from this in 1987. However, ISO 9001, as it is familiar to users today, only spread after the major revision of standard in 2000. Significant innovations were at that time a more comprehensible wording, more precise requirements and an improved applicability for service business. The strict process orientation is also attributable to this revision.

Today, the ISO 9000 series is regarded as the most important process-oriented standard worldwide. While ISO 9000 and ISO 9004 are explanatory and supportive, ISO 9001 is the only certifiable standard in this series. It is based on the idea that a QM system offers a transparent process for ensuring an appropriate quality level. The standard specifies minimum requirements that are independent of the specific service provision (product or service) and the size of the organisation in order to enable a uniform and comparable quality standard.

The certification according to the 9001 standard aims to,[2]

- create and maintain sustainable competitiveness through an effective QM system with efficient processes and their constant evaluation.
- constantly and systematically plan, implement, evaluate and improve the QM system.
- reduce the amount by which certified companies repeatedly deal with their own errors, weak points and waste in resources in order to eliminate the causes sustainably.

The implementation of an efficient QM system is seen as an overall operational task, which must start with all core processes. The main requirements of ISO 9001 therefore apply to the following areas:

- knowledge of internal and external issues as well as interested parties,
- responsibility and commitment of the management regarding quality policy and objectives, including the definition of responsibilities and authorities,
- establishment and maintenance of a process-oriented quality management system including knowledge and handling of operational risks,
- personnel qualification, operational knowledge, awareness and provision of resources including the associated documentation,
- recording and integration of customer requirements,

---

[2]Since ISO 9001 is not only suitable for companies, but also for public authorities, associations and other institutions, to use the term "organisation" instead of "company". According to the aviation standard, however, primarily private companies are certified, so that in the following text we will talk about companies.

- planning and execution of design and development (projects),
- selection, monitoring and control of external suppliers as well as evaluation and inspection of delivered products and services,
- planning and execution of the service provision including its release and post-delivery activities,
- process and product monitoring and measurement as well as analysis of the collected data,
- measures of corrective actions and risk minimisation as well as continuous improvement.

In terms of content, ISO 9001 (and thus also EN 9100) remains largely unspecific. Although system standards define what must be implemented in the end, they do not define *how* processes and tasks must be designed in detail. No tools, instruments or implementation methods are specified, only the requirements for outputs are defined. Quality management system standards leave the detailed process design, i.e. the choice of means, to the companies.

A QM system certification is accompanied by some disadvantages, as it is not the product or service quality that is checked, but the structure and organisation of processes of a company. This is often not sufficient to meet the wide range of quality requirements of many large multinational corporations, which therefore place their own requirements on their suppliers independently of standards. In addition, the quality requirements of ISO 9001 are not excessively high and thus even companies without a sustainable quality awareness can obtain the corresponding certificate.

1.2   High-Level Structure

All management system standards have a uniform design, the so-called *high-level structure*. This means that most chapters and top-level sections are identical to that of all important system standards. ISO 9001, EN 9100, IATF 16949 (Automotive), ISO 14001 (Environment), OHSAS 18001 (Occupational Safety) and ISO/IEC 27001 (Information Technology) as well as other standards have a common structure of their chapters in accordance with Fig. 1.1. This is reflected by the alignment of the contents of the standards with the terminology.

The high-level structure makes it easier for companies and auditors to work with multiple certifications because it simplifies a consolidated presentation of their own quality management. Different standards can be linked to one and another within a company. They do not have to be dealt with individually. without following up on different standards simultaneously. However, there is no obligation for the companies to adapt the high-level structure for their own QM system if only the respective standard requirements are to be fulfilled.

4 Context of the organisation
4.1 Understanding the organisation and its context
4.2 Understanding the needs and expectations of interested parties
4.3 Determining the scope of the QMS
4.4 XYZ management systems (standard-specific)
5 Leadership
5.1 Leadership and commitment
5.2 Policy
5.3 Organisational roles, responsibilities and authorities
6 Planning
6.1 Actions to Address Risks and Opportunities
6.2 Quality objectives and planning to achieve them
7 Support
7.1 Resources
7.2 Competence
7.3 Awareness
7.4 Communication
7.5 Documented information
8 Operation
8.1 Operational Planning and XYZ (standard-specific)
9 Performance evaluation
9.1 Monitoring, measurement, analysis and evaluation
9.2 Internal audit
9.3 Management review
10 Improvement
10.1 General
10.2 Nonconformity and corrective action

Fig. 1.1   High level structure for ISO management systems

1.3   Fundamentals of the EN 9100

Based on the ISO 9001, several industry-specific standards were developed at the end of the 1990s in which supplementary requirements of the respective industries were considered. In addition to EN 9100 for the aviation industry, ISO/TS 16949 for automotive engineering and TL 9000 for telecommunications have also emerged. These niche standards mostly arose from quality agreements that dominant market participants (e.g. Airbus, Boeing, Telekom and car manufacturers) demanded from their suppliers. The development was favoured by the fact that, based on such individual agreements, industry associations also issued quality standards in parallel or in addition to ISO 9001. Long before the first publication of EN 9100, the Airbus quality specifications in the 1990s had a decisive influence on quality standards published by national aviation associates in Europe. In addition, the publication of the American AS 9100, which is equivalent to EN 9100, had considerable influence on the publication of a separate aviation standard on a European level just before the end of the last millennium.[3] As a direct

---

[3]As a result, the European Association of Aerospace Equipment Manufacturers (AECMA) was mandated by the European Committee for Standardisation (CEN) to develop European Standards (EN) for the aerospace industry.

consequence, EN 9100 was published in 2003 by the European Committee for Standardisation (CEN) as the first certifiable standard for aviation, space and defence organisations. In 2005, EN 9110 followed for maintenance companies and EN 9120 for distributors. In 2009/2010 and 2016/2017, all three aviation standards were again substantially revised.

The International Aerospace Quality Group (IAQG) and the European Aerospace Quality Group (EAQG) are in charge of the further development of the EN 9100 series and represent European interests through their support.

EN 9100:2018[4] Contains the Complete ISO 9001:2015
The supplementary requirements of the aerospace industry in the 9100 standard text are shown in bold and italic letters and can thus be clearly distinguished from the classic ISO 9001 elements. Essential additions to EN 9100 compared to ISO 9001 are for instance:

- configuration Management,
- product safety requirements,
- requirements for handling counterfeit parts,
- the dedicated handling of special processes and critical items,
- more detailed requirements for supplier monitoring and
- further requirements for operational risk management,
- higher requirements for verification and validation,
- Process measurement and monitoring of the achievement of objectives via the so-called PEAR forms.

These extensions bring the EN closer to the EASA regulations (especially the Implementing Rules of Part 21 and 145), although considerable differences remain. While the EN focuses above all on customer satisfaction and process orientation, the EASA regulations focus on the safety aspect. In this respect, it is not surprising that both Airbus and the 1-tier suppliers, i.e. the Airbus' direct first level suppliers, normally require their suppliers to provide proof of EN certification.

This constraint for certification means that the suppliers themselves are responsible for proving their quality capability. They must assign accredited certification bodies at regular intervals in order to have their own EN standard conformity checked and confirmed. The certification issued on this basis then serves as proof to the supplier's customers. Thereby the OEMs themselves demonstrate the quality capability of their suppliers to their aviation authorities or their own customers. At the same time, customers can reduce their costs, especially for on-site monitoring through supplier audits. The advantage for the corporations is that they partially outsource their supplier monitoring.

Especially for suppliers of the lower levels of the supply cascade, certification does not necessarily imply additional cost. Many companies, especially the

---

[4]In the following only briefly EN 9100.

smaller ones, are for the first time systematically addressing the issues of quality management and process orientation as part of EN 9100 certification. The standard can therefore help to improve the operational value creation as well as interfaces to the customer. In this respect, certified companies often have a more pronounced process and quality awareness.

EN 9100 certification is also useful for those companies that are seeking approval under aviation legislation (production, maintenance, design). In this case, an accepted quality management system can be used which already adheres to the official requirements in many respects.

1.4   Fundamentals of EN 9210

In addition to the EN 9100, there is the EN 9120 for distributors in the aerospace and defence industry and the EN 9110 for aviation maintenance organisations. The EN 9110 will not be examined here as only approx. 250 companies, mainly corporate groups, are certified according to this standard in Europe.

A characteristic feature of the EN 9120 is its focus on the special requirements of the industry of distributers and stockholder. At the same time, EN 9120 does not contain all the requirements for production organisations. As of 2019, there are about 800 operating sites in Europe that are certified according to EN 9120:2018, and almost 2000 worldwide.

EN 9120:2018 also contains EN ISO 9001:2015 in its entirety. The supplementary requirements of the aerospace industry are shown in bold type and italics in the text of the standard and can thus be clearly distinguished from the classic ISO 9001 elements. Compared to the "normal" aviation standard EN 9100, EN 9120 mainly contains these supplementary requirements:

- electronic documents and records proving origin (Sect. 7.5.3),
- handling suspected unapproved parts (Sect. 8.1.5),
- storage and delivery, especially for split products (Sect. 8.5.2),
- dealing with nonconformities (Subsect. 8.7).

Compared to the reference EN 9100, the standard for distributors does not include requirements in the following areas:

- parts of production planning and control (Subsects. 8.1 and 8.5),
- risk management (Sect. 8.1.1),
- product safety (Sect. 8.1.3),
- validation of special processes (Sect. 8.5.1.2),
- First Article Inspection (FAI) (Sect. 8.5.1.3),
- usually also on design/development (Subsect. 8.3).

In order to be certified according to the distributor standard EN 9120, the share of the added operational value in aviation, aerospace or defence is not relevant.

Instead, it is compulsory that the company to be certified has the characteristics of a distributor or stockholder and does not simultaneously carry out processing activities. While separating batches, cutting to size and preservation are permitted, even the smallest manufacturing activities (e.g. assembly) lead to an exclusion from EN 9120 and to a progression to EN 9100 or EN 9110.

The EN standard EN 9120 for distributors and stockholders can currently be certified in the Revision 2018. This is equivalent to the American AS 9120 and the Asian JISQ 9120.

In the following chapters, essential differences to EN 9100 are shown in the text, smaller differences in footnotes.

Reference

Hinsch, M.: Industrial Aviation Management: A Primer in European Design, Production and Maintenance Organisations. Heidelberg/Berlin (2019)

Key Characteristics of EN 9100:2018

2

2.1   Process Orientation

Since its major revision in 2000, ISO 9001 has followed the approach of process-oriented quality management, which was adopted by the EN 9100 with its publication in 2003. The corresponding requirements were tightened in its first revision in 2009 and again in 2015 with the publication of EN 9101:2018[1]. The process orientation is therefore also reflected in the structure of the standard itself. For EN certification, an expanded understanding of process-based business organisation is therefore essential.

Process orientation is characterised by moving from a primarily department-orientation of service provision towards a procedural systematisation. The documentation of the processes makes an important contribution to this. To this end, the organisation must be divided into key-/core processes as well as management and support processes. These must be identified initially, then managed and finally monitored. The focus must not only be on the processes themselves, but also on their interactions, interfaces and performance measurements.

Through this approach, the process orientation demands and promotes a stronger examination of operational processes and responsibilities. The organisation is made more transparent and thus facilitates the clarity of the process structures. The employees recognize their role within the processes relevant to them as well as their role within the entire value chain.

For the success of the process-oriented approach and thus also for passing of the certification audit, it is important that an internal control loop exists between the incoming customer requirements (input) and the customer satisfaction (indirect output). The EN 9100:2018 requires the implementation of Deming's PDCA cycle

---

[1]EN 9101:2018 defines the requirements for the preparation, execution and documentation of certification audits in the aerospace industry.

© Springer-Verlag GmbH Germany, part of Springer Nature 2020
M. Hinsch, *Guideline for EN 9100:2018*,
https://doi.org/10.1007/978-3-662-61367-2_2

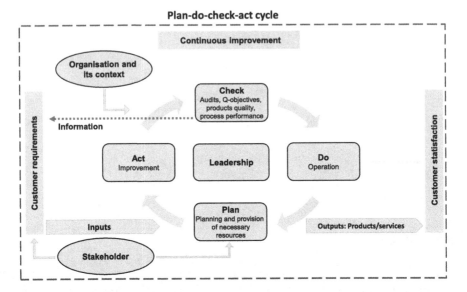

Fig. 2.1 PDCA cycle (following ISO 9001:2015, chap. 0.3)

(Plan-Do-Check-Act) (see Fig. 2.1)[2]. Consequently, the input of the customer, the requirements of the relevant interested parties, the context of the organisation and the operational resource management (*Plan*) form the input for the service provision. The value-added process (*Do*) and its output are subject to monitoring, performance measurements and analysis, product conformity and customer satisfaction (*Check*). Based on the results of this monitoring, management improvement measures must be derived, if planned results are not achieved in order to improve future service provision. Effective necessary actions shall be evaluated (*Act*). The entire cycle is subject to systematic leadership. The PDCA cycle must not only be reflected in the value creation process, but also, for example, in staff qualification and risk management as well as in the process for handling of non-conformities. All process specifications shall always take the PDCA cycle into account.

Process Documentation
The type and scope of process documentation depends on the individual operational conditions. Methodically, however, only a visually anchored organisational and process concept can create sufficient transparency.

At the highest level, a process map is to be defined (cf. Fig. 4.1) in order to obtain a complete overview of the company and its core processes. On the second level, which serves to describe individual processes, flow charts, flow diagrams or

---

[2]See also Fig. 2.2 in Sect. 0.3.

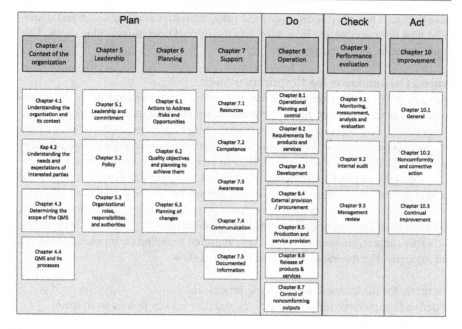

Plan				Do	Check	Act
Chapter 4 Context of the organization	Chapter 5 Leadership	Chapter 6 Planning	Chapter 7 Support	Chapter 8 Operation	Chapter 9 Performance evaluation	Chapter 10 Improvement
Chapter 4.1 Understanding the organization and its context	Chapter 5.1 Leadership and commitment	Chapter 6.1 Actions to Address Risks and Opportunities	Chapter 7.1 Resources	Chapter 8.1 Operational Planning and control	Chapter 9.1 Monitoring, measurement, analysis and evaluation	Chapter 10.1 General
Kap 4.2 Understanding the needs and expectations of Interested parties	Chapter 5.2 Policy	Chapter 6.2 Quality objectives and planning to achieve them	Chapter 7.2 Competence	Chapter 8.2 Requirements for products and services	Chapter 9.2 Internal audit	Chapter 10.2 Noncomformity and corrective action
Chapter 4.3 Determining the scope of the QMS	Chapter 5.3 Organizational roles, responsibilities and authorities	Chapter 6.3 Planning of changes	Chapter 7.3 Awareness	Chapter 8.3 Development	Chapter 9.3 Management review	Chapter 10.3 Continual Improvement
Chapter 4.4 QMS and its processes			Chapter 7.4 Communication	Chapter 8.4 External provision / procurement		
			Chapter 7.5 Documented Information	Chapter 8.5 Production and service provision		
				Chapter 8.6 Release of products & services		
				Chapter 8.7 Control of noncomforming outputs		

Fig. 2.2 PDCA structure of the EN 9100:2018

turtle diagrams (*Turtles* see Fig. 11.2) are used. Tasks, processes and procedures, which were in a function-oriented approach summarised in writing, are visually shown here in process flow charts. (cf. e.g. Fig. 7.1). The interactions between processes can be displayed, for example by using arrows.

Only the third level contains additional written instructions to the visualisations, as they are partly known from old procedural instructions. Through this multi-level structure, a process-oriented QM system creates transparency and thus emphasises on its advantages compared to written documentation describing the functions. These are:

- The visualisation is analogous to the natural value creation process,
- the focus is not on the hierarchy or departmental thinking, but on the process result,
- the multi-level process structure (process maps, processes, activities) increases comprehensibility for the employee,
- formerly isolated documentations are replaced by the sequencing of individual process steps with process flow orientation,
- due to its clarity and clear structuring, this methodology is well suited for the induction of employees and as an instrument of inhouse training.

Although the process-oriented approach in QM documentation is therefore very user-friendly, employees must nevertheless be introduced to this form of presentation. They must rediscover their roles, activities and interfaces and understand how their actions are integrated into the overall operational value chain. This should be reflected in training plans.

Process Orientation in the Certification Audit

With EN 9100, process orientation not only plays an important role in the operational QM documentation. Since the standard was revised in 2009, certified organisations have also been subject to a stronger obligation to provide evidence of the key process performance. Finally, the process-oriented approach was intensified in the course of the last revision of EN 9101 in summer 2015. From thereon, the certification auditors must audit the key processes in a binding manner and align their records with them. Also, the scope for the assessment of the process effectiveness for the auditor has been defined more precisely. Companies can no longer work without a clear process orientation of their organisation.

In this context, companies are not only required to define their processes clearly and logically. For the audit, they must be able to show

1. criteria for the evaluation of the core processes,
2. defined measurable process objectives and derived key figures from them,
3. measures to achieve the process objectives and
4. that the process performance is regularly measured and compared it with the process objectives.

This must be used to develop a cycle that recurs at least once a year. The aim must be to achieve a continuous improvement process with the help of the key figures. The certification auditor will therefore analyse the key figures collected in each audit, evaluate their development and compare them with the planned achievement of objectives. In this respect, even the smallest company must have defined at least two key figures with a target value for each key process, such as design and production, as well as for procurement. The larger the company, the higher the expectations regarding the scope and quality of process measurements and target tracking. The exact requirements for operational process measurements depend on the product portfolio and the type of service provided. The certification auditor assesses whether a system of process measurements is sufficient. Further information on determination of targets and objectives is given in Sect. 9.1.1 of this book.

In the certification audit, the evaluation of the process performance is carried out using so-called PEAR forms[3]. In this report, the auditor must describe the process in detail, document audit observations as well as the method and results of the process measurements and evaluate the effectiveness of the process. Deviations from planned targets shall be subject to an audit finding unless appropriate action has been taken.

---

[3]PEAR = Process Effectiveness Assessment Report. The form is Form 3 of EN 9101:2018, which can also be downloaded from the IQAG website (http://www.sae.org/iaqg/forms/index.htm).

2.2   Risk-Based Thinking and Risk Management

According to ISO 9001:2015 Subsect. 6.1 risk-based thinking and actions are required[4]. Therefore, a fundamental risk orientation is not new for the EN 9100, since the aviation standard has its own chapter on risk management since 2009. Overall, however, the importance of risk orientation has increased considerably with the last revision of the standard. This is made clear by the fact that the concept of risk is mentioned 60 times in 17 (sub)chapters.

Thereby, there is a strict separation between overall operational risk on one side and operational risks on the other. Consequently, they are treated differently. Thus, the standard requires

- a risk management for dealing with operational risks within the scope of service provision, especially for the processing of orders or projects (cf. Sect. 8.1.1).
- risk-based thinking to focus on the complete company (to a lesser degree than risk management), according to its QM system and its processes (cf. Subsect. 6.1).

In both cases, the aim is a targeted confrontation of the operational risks. One of the main tasks is a timely identification of risks and to take targeted actions to keep them under control or eliminate them wherever possible.

Detailed information on risk-based thinking and risk management can be found in book Subsect. 6.1 and Sect. 8.1.1.

2.3   Customer Focus

Customer orientation is a key characteristic not only in numerous business management approaches, but also in the EN 9100. The customer should therefore be at the centre of the company's activities.

A consistent process orientation is required for a successful customer orientation. Today's customers' basic needs such as flexibility, on-time delivery, short reaction times and low prices, can only be met if the company's own operational processes are properly coordinated and connected to the customer's requirements. The standard requires the implementation and monitoring of customer orientation in numerous chapters. This means that the customer focus must be anchored operationally and strategically in the company. Questions are for example:

- What does the customer want? What obvious requirements does he formulate?
- What expectations does the customer have that are put in writing?

---

[4]See EN 9100:2018 Subsect. 6.1.

- Regarding the product or service.
- Regarding the speed or reliability. How does the customer define these aspects?
• How is it ensured that customer's requirements are systematically identified and met within the organisations own processes?
• How are employees informed about customer's requirements?
  - How is it ensured that employees are aware of these requirements in the context of their own work?
  - Are employees aware of what non-compliance with customer requirements within their area of responsibility means?

At the company level, EN 9100 requires the collection of parameters that provide indications of customer satisfaction. To this end, the standard in Sects. 5.1.2 and 9.1.2 makes the measurement of the following parameters mandatory:

• product conformity,
• punctual delivery (On-Time-Delivery),
• customer complaints,
• requests for corrections (rework).

For these parameters, key performance indicators (KPIs) must be available for every certification audit. This is not negotiable!

At the operational level or project level, Subsect. 8.2 explicitly focuses on customer orientation with regards to identification and evaluation of customer requirements. In particular, effective procedures must be set in place when initiating orders. For this purpose, inquiries, quotations, contracts and their modifications must be documented in a comprehensible manner. This especially applies to complex products and services, where the offer is derived iteratively between customer and potential contractor. The use of CRM software[5], which is used for recording activities and customer communication in a structured manner, can be helpful. During the certification audit, the customer orientation during the quotation process is typically checked on random samples of records of completed orders.

References

CEN-CENELEC Management Centre: EN 9101:2018—Quality management systems—Audit requirements for aviation, space and defence organisations. Brussels (2018)
CEN-CENELEC Management Centre: EN 9100:2018—Quality management systems—Requirements for aviation, space and defence organisations. Brussels (2018)

---

[5]CRM = Customer Relationship Management. This type of software serves to manage customer interaction so that, in addition to offers, orders and delivery data, email traffic, customer contacts, meetings and marketing campaigns can also be stored and thus made comprehensible.

Wording and Terms

3

At numerous points in the EN 9100, complicated and complex words are used which are difficult to understand when first read and which do not always make it easy for laypersons to develop an appropriate understanding of the requirements of the standard (see Table 3.1). The reason for this is that these are collective terms which must cover a wide range of applications.

It is not necessary to integrate such standard terms into one's own QM documentation or even to use them in day-to-day operations. It is recommended to omit these terms as employees with a low interest in the QM system also have to understand the QM language and the organisational QM documentation. Terms that are foreign to everyday life damage the company's QM acceptance in the eyes of its employees and should therefore be limited to the EN 9100 text.

There will be QM experts who, with or without justification, recognise fine differences between the EN terms and the synonyms of everyday business life. However, since the vast majority of QM users and 9100 auditors is not familiar to these differences, it is not to be expected that these fine differences will play a role in daily certification business.

In addition to the complex standard language in general, which is decoded as far as possible in chaps. 4–10, four additional terms which are of high significance for the EN 9100 are explained here[1]:

- counterfeit parts,
- critical items,
- key characteristics and
- special requirements.

---

[1]Cf. EN 9100:2018, Subsect. 3.1–3.5.

© Springer-Verlag GmbH Germany, part of Springer Nature 2020
M. Hinsch, *Guideline for EN 9100:2018*,
https://doi.org/10.1007/978-3-662-61367-2_3

Table 3.1 Examples of terms not typical for everyday life

Relevant interested parties (also: Stakeholders)	Persons or institutions (stakeholders) whose actions influence the provision of services of the company, e.g. end customers, suppliers, trade unions, citizens' initiatives, chambers and associations as well as competitors, investors or partners, but also think tanks or media
Documented information	Documents and records, videos, audio recordings, websites, files
External providers	Collective term for supplier, subcontractor, external company, affiliated companies, such as subsidiaries, sister companies or parent companies (outside the scope of certification)
External provisioning	procurement

Counterfeit Parts

Counterfeit parts are components, assemblies or materials that have been knowingly (!) manufactured and released in accordance with approved or recognised procedures. These may have been placed on the market with invalid, falsified or missing (release) certificates, accompanying documents, histories, or with incorrect markings or packaging. These parts therefore do not comply with the specifications, i.e. an approved type certification or the generally applicable norms or standards.

Critical Items

Critical items are those products, components or processes that have a significant and risk associated impact on the product realisation or operation of the product. Criticality can affect the safety performance, the "4 F" (form, fit, function, fatigue), the production process and quality, or product life. In addition to the critical items mentioned in Subsect. 3.2 of EN 9100, typical examples are components or systems that are critical to failure, items that are vital to the order regarding capacity or time, special processes, manufacturing processes which are particularly complexity and orders for which the necessary experience is lacking.

The control of critical items requires specialised control and monitoring measures in order to minimize risks during value creation or later operation. Such specific measures may include for instance:

- specific instructions for testing or checking critical items and specific procedures for their release (e.g. double inspections).
- specific information in the production specifications and/or job cards, e.g. by means of a prominent printed notice or a red stamp labelled 'critical item' or 'critical process' in order to increase the awareness of the production staff.
- employee training to ensure that staff have sufficient experience in handling of critical items and to improve the awareness of the particular risk of misconduct.
- subcontracting of the provision of services to specialised companies, the results of which must nevertheless be evaluated.

During a certification audit, it must be expected that the auditor will address critical items or identify such units. Since these are commonly examined more extensively, the organisation is ought to have established appropriate measures for their control.

Key Characteristic

A key characteristic is a product or process element whose modification has a significant impact on performance, product life, 4F or on production. Key features, as well as critical items, require special control and tracking.

Special Requirements

Special requirements are requirements whose fulfilment carries major risks. It is irrelevant whether these are identified by the organisation itself or by the customer. Special requirements must be identified and evaluated in the bidding process and are therefore explicitly mentioned in Sect. 8.2.2 (c) of the standards. By identifying possible risks before the first steps of the actual service provisions are taken, an awareness of possible risks should be created at an early stage. This is intended to ensure that appropriate measures are included in the planning and implementation of service provision.

Special requirements of customer orders may arise, for example, from

- highly complex production processes or,
- production processes for which the company lacks the necessary experience,
- production at the limits of what is technically or procedurally feasible,
- special processes,
- a contract-specific expansion of operational resources outside the usual framework,
- unfavourable working or environmental conditions,
- an unusual political environment or difficult market conditions.

Specific requirements may include the definition of critical items or the identification of key features. Conversely, the design output (cf. sect. 8.3.5) may result in critical items, which in turn may call for specific requirements regarding production, storage or transport.

Reference

CEN-CENELEC Management Centre: EN 9100:2018—Quality management systems—Requirements for aviation, space and defence organisations. Brussels (2018)

Context of the Organisation

4

4.1 Understanding the Organisation and Its Context

Subsection 4.1 contains the requirement to become aware of one's own position within the market and the operating environment as an enterprise. It is about answering operational questions beyond the day-to-day business with which every company is confronted:

- Where and how does the internal and external environment influence the operational service provision, the QM system and the quality objectives?
- What are the issues the company is currently dealing with or will have to deal with in the coming month and years? This means: which internal and external issues are relevant to the company beyond the daily operational business?
- What makes the company stand out on the market? What is the unique selling proposition (USP) of the company and in which direction is the portfolio to be expanded in the future?

Only if a company can provide answers to these questions it can achieve its long-term objectives. Many companies, especially small and medium-sized enterprises, lack a real strategic orientation because they already fail to clarify their long-term goal and what influence the market environment has on their operational success. Even if the management envisions the position of the company in the future, there is often a lack of a systematic and structured, traceable approach. Often this is caused by a reluctance towards changes. Even if such companies are currently successfully pursuing their strategy, this is generally not enough for the long term. In the EU, however, markets are increasingly characterised by displacement rather than growth. The ability and willingness to deal appropriately with internal and external influencing factors, is therefore decisive on the companies' survival in the long run.

© Springer-Verlag GmbH Germany, part of Springer Nature 2020
M. Hinsch, *Guideline for EN 9100:2018*,
https://doi.org/10.1007/978-3-662-61367-2_4

The starting point for a context analysis and for strategic positioning is the precise knowledge of where the company stands in the market environment. In order to determine a promising strategic orientation, it is therefore necessary to identify a target position. In this respect, the standard requires regular reflection on one's own internal situation and the external environment. Typical aspects for an evaluation of the external environment are the market strategy of competitors, customer demands, adaptation of the product portfolio, expansion of operations, innovations and technical developments, effects of digitisation or the changes to personnel, legislative initiatives, activities of chambers and associations, etc.

Solid methods for determining the context of the organisation are, for example, systematic strategy/long-term planning, PEST or portfolio research as well as GAP or SWOT analyses.

In the certification audit, the requirements regarding the type and scope of the auditor's assessment depends, as usual, on the individual circumstances. In any case, the major external and internal issues must be addressed during the management review (see Subsect. 9.3b).

The management has to demonstrate that the company is aware of its own operational strengths and weaknesses to the auditor. Furthermore, an understanding of the opportunities and risks of the market is to be revealed along with the companies systematic and compressible approach to deal with the market situation. In a 10-man operation, it may be sufficient for the manager to be able to present his business position convincingly by means of a brief documented overview, e.g. with the following statements:

- Employee Johnson is too unreliable to be a CNC-mechanic, the Shift leader constantly needs to check the task.
- An important competitor 30 km away has filed for bankruptcy. We are now looking for an interview with the insolvency administrator; under certain circumstances it may be possible to take over an employee, machines and orders.
- The large milling machine is 30 years old and must be replaced in two years at the latest.
- Some of our services are already being tested in 3D printing. We do not want to make these investments. Here we must define and acquire replacement business in the future. To this end, we are aiming to intensify our activities in the post-processing of 3D printing as an alternative.

These and other statements and records (e.g. management review) will generally be sufficient for small organisations. Finally, the main issues in the areas of market, competition, resources and legislation were presented. Since measures have also been outlined, appropriate awareness can also be assumed.

In corporate structures, however, such simple statements are not sufficient. Here, systematic and documented action must be demonstrated in discussions with the management, but sometimes also throughout the entire audit process. For example, the following plans or concepts shall be provided to give an impression of adequate environmental observation and organisational control:

- Corporate strategy,
- Financial, investment, personnel and project planning,
- Market, competition and SWOT analyses,
- Risk activities/management,
- Observation of economic aggregates (e.g. interest rates, exchange rates),
- Benchmarking,
- Make-or-buy analyses,
- Product development reports and,
- Meeting minutes, task lists and effectiveness checks.

4.2   Understanding the Needs and Expectations of Interested Parties

Companies must not only deal with the question of *what* their influencing factors are, but also *who* is influencing the organisation's development. Such interested parties or stakeholders are all those institutions, groups or individuals that have a direct or indirect influence on the performance of the company.

Interested parties are for example direct or indirect customers, suppliers, employees, trade unions, associations, initiatives or chambers as well as competitors, investors and partners or media.

However, the focus must explicitly not be on all interested parties, but only on those that are "relevant" from the management's point of view. It is therefore not a question of fulfilling the declared or implied expectations of all stakeholders. The objective is, that companies should not appear isolated in their environment. They should be aware of their interested parties and their needs, analyse them and draw conclusions for their own operations and, if necessary (!), derive appropriate actions. The focus is thus on generating a continuous awareness for the viewpoints, requirements and needs of the market participants.

A fundamental and documented description of the interested parties must be available for the audit (cf. sect. 4.4.2), for instance in the form of an Excel spreadsheet or a mind map which briefly describes the type of influence or current issues or requirements as well as risks and opportunities of the stakeholder on their own business.

In the certification audit of the previously mentioned 10-man operation, it will generally be sufficient for the manager to be aware of the following interested parties, for example, and to be able to express their essential requirements and possible risks and opportunities:

- main customers and their end customers,
- A-suppliers,
- employees,
- bank,
- competitors,
- municipality,
- landlord, if applicable.

Airbus or large Tier-1 suppliers, at the upper end of the value chain define the other end of the spectrum, as they are confronted with countless interested parties due to their large company size. Here, the requirements, opportunities and risks as well as measures for the following groups must be recorded and coordinated:

- customers and customer groups (airlines, maintenance organisations),
- indirect customers (passengers, grouped by first, business and economy, freight companies, airports),
- suppliers,
- politics (local, state, federal, EU),
- owner/shareholders,
- local, national and international or EU authorities (environmental or safety authorities, Federal Aviation Authority, EASA, FAA, TSA, ICAO),
- citizens' initiatives,
- associations and clubs (EAQG, Greenpeace, Trade unions),
- Employees, potential employees.

In large companies it is usually necessary that there are "caretakers", i.e. designees or departments that systematically deal with the concerns of these interested parties and that can be questioned in a certification audit.

4.3   Determining the Scope of the Quality Management System

The scope of the QM system must be defined and recorded in writing. It must be defined for

- the location(s) of the company,
- the range of products and services,
- partial certifications or the scope of application within a site.

The EN 9100 is a standard that must be applicable for a multinational corporation, for a medium-sized component manufacturer, but also for a small engineering service provider or a 3-man operation for software development. Under these conditions, compromises are necessary and not all requirements or chapters can be applied to every company in the aerospace industry. For example, a built-to-print subcontractor carrying out grinding work for a component manufacturer will usually not carry out design-activities. In such cases, those requirements may be excluded from the application or declared as inapplicable.

However, non-applications are only permitted if the requirements of the standard to be excluded do not affect the QM system, customer satisfaction or product or service conformity. Therefore, non-applicabilities of everyday certification practice is usually limited to the area of product and service realisation (Chap. 8

or Sect. 7.1.5)[1]. Finally, all other chapters of the standard have an influence on the QM system and there cannot excluded normally.

Any non-applicability of requirements shall be justified and documented. The QM manual is recommended as a documentation medium.

4.4   Quality Management System and Its Processes

Subsection 4.4 deals with fundamental requirements for the QM system and requirements for process orientation.

At the beginning of Subsect. 4.4, a requirement for the permanent presence of an effective QM system is formulated. It is not worth taking a closer look at this basic requirement at this point, as it is specified in more detailed in all subsequent EN-chapters. If requirements are applicable, they must be fulfilled. If individual standard requirements conflict with legal or official requirements, the latter apply in accordance with Chap. 1 of EN 9100:2018.

The main part of Subsect. 4.4 is dedicated to the operational process orientation. With the corresponding requirements it is intended that the provision of services is primarily aligned with the ideal process flow and is not determined solely by the functional organisational structure (hierarchy). The aim is to achieve a stronger orientation of value creation towards the needs of the customer. This is because the functional alignment according the organisational chart often favors individual departmental interests over an increase in customer satisfaction. In contrast, the process-oriented organisational structure is more resource-efficient and geared stronger to customer needs.

For the certification audit, it is necessary to define a company process map (cf. Fig. 4.1). This must contain all required major processes for the provision of services. Certification is not possible without a process map, as the outline of the audit and process measurements follow this map.

The listing in Sect. 4.4.1 formulates more detailed requirements for the implementation of process orientation. Numerous specifications are formulated in more detail elsewhere in the standard (and are therefore not further explained here):

(a) Inputs and outputs of the processes must be defined. This information must be available in the process descriptions. Alternatively, it is possible to use process turtles (turtle diagrams). Following is an example for a manufacturing process:
    (a) Input: Material, purchased parts and job cards from planning.
    (b) Output: component approved by QA, stamped job cards, test report
(b) It is not enough to define the operational processes on their own. The relation between the processes must also be defined. These interactions can, for example, be determined by the IT structure by the means of a workflow, by written

---

[1]Note: Sect. 8.5.5 *Activities after delivery cannot be* excluded.

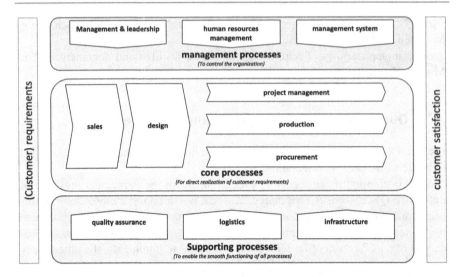

Fig. 4.1 Example of an operational process map

process descriptions or by verbally expressed best practices. The relation between the core processes and important support processes and management processes must be mapped using a process map (cf. Fig. 4.1). At the process level, interactions in process descriptions can be visualised through adjacent processes and, if necessary, described in writing.

(c) It is necessary to measure process performance. The standard requires all processes that are necessary for the functioning of the QM system to be monitored using meaningful key performance indicators (KPIs). This means that key figures must not only be collected for the core processes, but also, for example, for planning processes as well as correction and improvement processes. (See also the excursion *Process measurement, KPIs and data collection* at the end of this chapter).

(d) In the certification audit, core processes are monitored using so-called PEARs (Process Effectiveness Assessment Reports). The company must inform the auditor of the associated KPIs in the audit. Therefore, these KPIs are to be collected and evaluated on a monthly basis.

(e) Also, for all other processes, a performance measurement and evaluation must be ensured. The auditor will therefore ask: *How do you determine the effectiveness of the key-processes and what is the current performance status?* This is checked during every certification audit, although usually not stringently:

(f) see also Sect. 7.1.1 and Subsect. 8.1

(g) see also Subsect. 5.3 b

(h) see also Subsect. 6.1

(i) see above, bullet point c

(j) see above, bullet point c and f.

In Sect. 4.4.2 some requirements regarding documented information are formulated. This documentation must be created, utilized and stored.

In the aerospace-specific part of this subchapter (bold and italic text) several extended mandatory documentation requirements are listed:

- a description of the relevant interested parties (according to Subsect. 4.2),
- the scope of the QM system, including restrictions and non-applicabilities (according to Subsect. 4.3),
- the sequence and interaction of these processes (see Sect. 4.4.1 b)
- a designation of the responsibilities and authorities for the processes of the QM system (see Subsect. 5.3 b).

Although a QM manual is not explicitly required in formal terms (anymore), there are overall documentation requirements, which can be integrated efficiently in the QM manual. In this regard, the QM manual is a useful document for most companies. A specific structure of the QM manual is not required.

Excursion: Process Measurement, KPIs and Data Collection
Regarding processes monitoring, there is often uncertainty, especially in companies pending their first certification, as to which KPIs (Key Performance Indicators) are suitable or which are accepted by the certification auditors. It is difficult to provide a rule of thumb for this. Process monitoring should help the management to identify weak points in and to initiate improvement measures according to the PDCA-Cycle. With the monitoring and measurement activities, the management must be able to control the operation. In this respect, KPIs are a tool for deriving decisions based on solid facts.

Fundamental key figures with a direct relation to the defined core processes are required. These are to be monitored and controlled with concrete (annual) targets. In case of deviations from the target or thresholds, countermeasures must be taken systematically. Methodically, it is therefore important that the objectives are SMART, i.e. **s**pecific, **m**easurable, **a**cceptable, **r**ealistic and **t**erminable. At least two KPIs must be defined for each core process. *This* is not negotiable!

The type and scope of the key performance indicator observation are essentially based on the size of the organisation. The standard does not impose restrictions. As a guideline, however, about six to ten process KPIs (with annual target values) per 50 employees[2] should be an appropriate proof for a KPI-based process monitoring in a certification audit.

Key figures that are be collected in every case are product conformity/on-target-quality (OTQ), the on-time-delivery (OTD) and customer complaints. Financial data are not required by the norm[3].

---

[2]It should be possible to reduce the value from around 200 employees onwards.
[3]See EN 9101 Sects. 4.1.2.4 and 4.1.2.4 NOTE 2.

For the definition of key figures, the path of the operational process map must be followed. The first step is to ask what the output of each core process is, for example:

1. What is the role of the sales department?
2. When is the design process deemed successful?
3. What characterises is a good production?
4. When can the procurement of materials and services not be further improved?

In a second step, one or more answers must then be given primarily under aspects of quality and effectiveness, e.g.:

1. to acquire orders, i.e. to bring as many offers as possible to a successful conclusion.
2. If the expected product can be developed within the time, financial and resource constraints,
3. That the product was delivered without rework, without defects and without delay.
4. If the external procurement has been delivered on time, on quality and in accordance with all other requirements.

In order to answer these questions, it is necessary to precisely know the requirements and needs of external or internal customers. Only with this knowledge can the process owner define the right target values.

In a third step, these answers are used to derive KPIs, e.g.:

1. Hit rate or new customer turnover (i.e. quality of offers), determined by comparing the orders received in relation to the offers submitted,
2. Planning precision in the design department:
   (a) On-Time-Delivery: Deviation in days between planned milestone release and actual release, possibly in relation to the total length of the associated design phase,
   (b) Budget compliance: Actual costs to planned costs,
   (c) Compliance with planned project resources (hours or days): Actual hours booked on the design project to planned project hours,
3. Rejection rate in the final acceptance test, customer complaints (after delivery) or warranty costs,
4. Quota of goods receipt findings, On-Time-Delivery of vendors.

Depending on the size of the organisation and the service portfolio, the following KPIs can also be defined:

• Machine downtimes, alternatively machine utilisation,
• Waste, rejections, rework,
• Throughput and processing times,

- Waiting and idle times,
- Stamp rate of employees or booking rate on orders,
- Time span from incoming order to delivery,
- Error statistics of all kinds,
- Staff turnover rate,
- Duration of the processing of audit complaints,
- IT downtime,
- Duration of recruitment of new employees (from receipt of application to signing of contract),
- Material returns to stock,
- Cost of Non-Quality,
- Stock turnover and capital commitment,
- Storage temperature and humidity.

When defining ratios, it should always be noted that they should be easy to obtain and are unbiased. Finally, the process owner must be able to directly influence the values. All too often, controllers and quality managers collect data and provide KPIs that are of little use to the recipient. Before the survey is carried out, the requirements of the decision-makers should therefore be clearly identified. At the same time, it must also be clarified and documented who is responsible for what data, at what frequency and to whom it is to be made available, at what time and how.

With additional well-prepared visualisations for ongoing reporting, quality managers or controllers can score points not only in the context of the management review itself. Managers want information that is easy to understand. Even though visualisations alone are not enough in the certification audit, graphics, especially in the context of trend analysis, are often more helpful than numbers. By the integration of targets and warning limits, graphics can also quickly show when KPI values are still within the acceptable range or when they are approaching critical limits.

References

CEN-CENELEC Management Centre: EN 9100:2018—Quality management systems—Requirements for aviation, space and defence organisations. Brussels (2018)
CEN-CENELEC Management Centre: EN 9101:2018—Quality management systems—Audit requirements for aviation, space and defence organisations. Brussels (2018)
CEN-CENELEC Management Centre: EN 9120:2018—Quality Management Systems—Requirements for aviation, space and defence distributors. Brussels (2018)

Leadership

5

5.1 Leadership and Commitment

5.1.1 General Information

The Executive Board or the CEO is the ultimate management body of a company and has therefore a special responsibility for all business matters, which of course includes quality management. In this respect, management is responsible to establish, maintain and continuously improve an effective QM system that conforms to the EN 9100. This includes in particular the responsibility for ensuring the implementation of the following:

- Definition and communication of a quality policy and quality objectives,
- Leadership and responsibility towards employees and
- Implementation of a strict quality, process, risk and customer orientation,
- Establishment of repeatedly observable process flows,
- Definition of roles, responsibilities and authorisations,
- Provision of the necessary resources,
- Systematic monitoring and tracking of process results,
- Support of subordinate managers,
- Ensuring a system of continuous improvement.

Furthermore, the management influences the internal quality perception, the degree of implementation and thus the efficiency of the QM system with its actions. With the attitude of the management towards quality management, its success and acceptance in the entire organisation stands or falls. The standard therefore requires management to focus on creating motivation, understanding and

© Springer-Verlag GmbH Germany, part of Springer Nature 2020
M. Hinsch, *Guideline for EN 9100:2018*,
https://doi.org/10.1007/978-3-662-61367-2_5

awareness in addition to classical management tasks. In this sense, *leadership is* defined as the ability to motivate employees to always keep an eye on organisational objectives. Above all, this means that employees must understand what their respective tasks are and where management wants to take its staff. The goal is therefore to create a commitment between the management and the employees for this "journey". This requires, among other things, comprehensible communication, i.e. with respect to

- processes and their interactions,
- the importance and the tasks of an efficient QM system,
- quality policy and quality objectives (cf. also Sect. 5.2.2),
- effects of nonconforming provision of services and
- risk-based action.

With regard to the path to be taken, however, the text of the standard remains imprecise on the subject of leadership and thus does not formulate any expectations as to how this requirement ("showing leadership") can be met methodically. The *how* remains on a comparatively abstract descriptive level according to the listing a)—j) in Sect. 5.1.1. This is not unusual for management standards, but it makes it very difficult to formulate an audit finding on soft topics such as leadership. This becomes particularly clear in Sect. 5.1.1 h) and j) when it comes to concrete support for employees and managers. In addition, an audit finding regarding inappropriate leadership fails either because of the evidence or because of the sword of Damocles hanging over the certification body resulting in its loss of mandate.

5.1.2 Customer Focus

In addition to process orientation, a strict customer focus is an essential element of the EN 9100:2018, in the sense that the standard requirements and needs of the customer are systematically are identified, recorded, evaluated and finally taken into account during the product or service realisation. The company must also be in a position, to identify needs not mentioned by the customer and to incorporate them into the provision of services—after all, quality is primarily defined as the fulfilment of customer expectations. The know-how required for this can, for example, come from market knowledge or trend analyses, experience with or information from customers, interested parties or from previous orders.

In order to ensure long-term customer orientation, the standard references not only the fulfilment of product and service requirements but also the risk management and pursuit of opportunities. These obligations are formulated in more detail elsewhere in the standard (Sect. 8.2.3 and Subsect. 6.1 and Chap. 10) and therefore are more of a memorandum than an instruction for actions.

This can be implemented in everyday business life, e.g. through proactive customer communication and identification of customer needs that are not explicitly mentioned. Other ways to demonstrate customer focus are internal communication of set objectives regarding customer interaction, management of customer-related risks or active provision of product updates.

The measurement of customer satisfaction must be carried out for the certification audit in accordance with Sect. 5.1.2 d) by regular identification of the

- On-Time Delivery (OTD),
- Product and service conformity (On-Target-Quality—OTQ)

Other sources for determining customer satisfaction include the measurement of customer complaints, warranty claims or customer surveys using structured discussions or interviews. In addition, customer satisfaction can also be determined through issues from visit reports or telephone calls with customers. These can then be evaluated, and measures derived through regular brainstorming sessions.

Further information on customer orientation can be found in Subsect. 2.3 of this book.

5.2   Quality Policy

The quality policy describes the quality guidelines and the quality standards of the company. These are principles that show how the management sees or would like to see the company positioned. When formulating the quality policy, it is not required to stress quality, process or customer orientation explicitly in every sentence. However, it must be clear that quality is of great importance to the management. The overall quality policy must be coherent. At the same time, the policy should be formulated in such a way that the quality objectives can be aligned to the base.

The quality policy includes an obligation of the management to

- consider relevant requirements, especially those of customers, authorities and legislators,
- continuously improve the QM system.

To ensure that the quality policy has a practical benefit, it must be communicated to the employees, understood and lived. Thereby the quality policy can be used as a management tool. It is therefore important that the staff understand where the management wants to take them and what the quality policy means for them and their work. In order for the quality requirements to be credible, the management must of course set an example.

The quality policy must be documented—where and in which form is not defined by the standard. If the QM manual is still used for documentation purposes, it is advisable to supplement this with an announcement by means of an e-mail circular or by posting it on the bulletin board or framed next to it, which then underlines its value. Whether kitchen, info board, entrance area or bathroom—it is important to find a place where the employees perceive the quality policy. Therefore, documentation in the QM manual alone is usually not effective.

The quality policy must be reviewed regularly, i.e. at least once a year, e.g. in the course of the management review (cf. Subsect. 9.3) and adjusted if necessary.

The quality policy of most companies usually has an unspecific, sometimes visionary character, so that a strategy must be added. While quality policy formulates orientation and priorities, the strategy goes further and contains a rough path and instruments for the implementation of the quality expectations. A strategy as a step between policy on one hand and quality objectives on the other is not explicitly mentioned in EN 9100. However, management is obliged to plan appropriately so that long-term (strategic) organisational control is necessary (e.g. annual or budget planning). The strategy is usually fixed in writing, although it may be sufficient for small and medium-sized enterprises (SMEs) for the manager to verbally explain the strategy during the certification audit and to provide objective evidence of the path taken.

According to Subsect. 5.2 c), the quality policy must also be made available to the relevant interested parties (upon request). However, the wording is so soft ("*relevant* interested parties" and "as far as appropriate") that a company which does not wish to meet this requirement will usually find a justification for not sharing.

In a certification audit, the auditor will usually not only examine the nature and appropriateness of the quality policy by means of documented information and a brief discussion with management but also by determining whether the employees are aware of the quality policy (cf. also Subsect. 7.3). For this it is not mandatory to recite the policy, but it is necessary to explain the contents of the policy in one's own words. Even more important is that employees have an idea of what the quality policy means for their own range of tasks. The connection between daily work and quality policy must therefore be known and, if necessary, briefly explained in the audit.

Example for a Quality Policy and Declaration of Commitment

The management of Example-Component Ltd. regards quality and customer orientation as a strategic corporate goal. We oblige the following quality principles, it is our duty to distribute these corporate principles, to set an example and to derive goals from them:

- We want to grow continuously. With our high-quality products, we not only want to serve European customers, but also increasingly convince those of the American market over the next few years.
- Innovations determine our future. With our products and services, we provide cutting edge technology which we strive to continuously improve.
- In our market segment, we want to be regarded as a premium provider. This is only possible if we deliver our products in the expected and required quality. The fulfilment of the quality requirements and the individual expectations of our customers as well as compliance with legal and regulatory requirements or those of interested parties are therefore our performance benchmark.
- We expect the same quality standards from our suppliers as we expect from ourselves.
- We only make an error once. Constant improvement and detailed root-cause analyses are therefore an important topic for us.

For the implementation of these quality guidelines, a quality management system in accordance with EN 9100 has been implemented in our company, which is applied to all operational areas. This is intended to help our managers and employees to meet customer requirements at all times and to permanently ensure compliance with legal and regulatory requirements. Here they can count on the support of our Quality Management Representative (QMR), who ensures that the QM system is adhered to, lived and constantly improved by all employees. For this the QMR receives the full support of the management.

London, May 2020

Peter Clark

Managing Director Example-Component Ltd.

5.3   Organisational Roles, Responsibilities and Authorities

An orderly provision of services is only possible if the responsibilities are clearly defined. Therefore, companies must define responsibilities and authorisations for their jobs, positions and designees. The definitions necessary for this are to be documented in the organisation chart, in job descriptions and in process instructions as well as in procedures and the QM manual.

The responsibilities and authorities must be communicated and understood within the company. Every employee must know his or her area of responsibility. To prove this in an audit, a copy of the current job description signed by the employee should be archived in the personnel file (*"I have read and understood the responsibilities and authorisations defined in this job description"*). It is not mandatory be the standard, but it helps strengthen the awareness for the own responsibilities and authorisations. This can be of importance not only for EN certification, but also in the context of releasing managers from liability in the event of occupational accidents or in case of misconduct and its legal consequences.

It is in the range of tasks of the management, to define responsibilities and authorisations for essential QM activities. QM responsibility is therefore not limited to the QM representative, because in daily practice not only the QM representative is responsible for achieving appropriate quality and customer orientation. After all, every employee is responsible for quality and carries the responsibility for their own work.

While the ISO 9001:2015 does not require a QMR, the position of such a management representative is mandatory according to EN 9100:2018. This person is responsible for coordinating quality management tasks and ensuring the operational fulfilment of the EN 9100 requirements. The QMR is the person of contact and provides support for all quality management issues in all areas of the company. The QMR collects and prepares data to enable analysis and evaluation of the process performance and the effectiveness of the QM system. The QMR informs the management about the current status of the QM system, including the need for improvement, by means of regular reports and management reviews.

The QMR must be an employee of the company. Merely an external consultant is therefore not acceptable as a substitute for an internal QMR. However, an external consultant may provide significant input to the internal QMR.

The QMR must have organisational independence and direct access to top management. In the organisation chart, it is therefore important to ensure a direct link between top management and the quality management representative (QMR). Nevertheless, a QMR can also perform other operational tasks at the same time, e.g. as a buyer or a work planner. However, it must then be clear that the holder of the QMR position has two operational roles and may also be subordinate to two direct superiors. In the above-mentioned example, this would be the purchasing manager or the production manager on one hand and the CEO for the QMR function on the other. If the QMR holds more than just this role, it must be noted that he or she is not allowed to carry out internal audits in the QM department as well as in the area of his or her second or main activity.

As a rule of thumb, a company should provide one full-time position for quality management per 100 employees.

The standard does not require any specific qualification of the QMR. However, for carrying out this role properly, a basic understanding of quality in general and the EN 9100:2018 in particular is necessary. Therefore, it is highly recommended that the QMR has attended an EN 9100 basic training and an auditor course before starting work. In addition, a QMR should have a certain affinity to quality management, as otherwise there is a danger that QM activities will be limited to a minimum.

References

CEN-CENELEC Management Centre: EN 9100:2018—Quality management systems—Requirements for aviation, space and defence organisations. Brussels (2018)
CEN-CENELEC Management Centre: EN 9120:2018—Quality Management Systems—Requirements for aviation, space and defence distributors. Brussels (2018)

Planning

<big>**6**</big>

6.1 Actions to Address with Risks and Opportunities

Every company is obliged to manage its own operational risks and opportunities—continuously and proactively. Companies must anticipate their risks, assess their influence and deal with them appropriately. A distinction is made by the EN 9100 between two different sources of risk, each of which has its own requirements:

- General organisational risks or risks with direct effect on the QM system,
- Risks on orders or project levels.

While this Subsect. 6.1 is dedicated to the requirements of operational and QM risks, Sect. 8.1.1 focuses on order and project risks.

EN 9100:2018 requires a risk-based approach for risks that affect the entire organisation or the QM system systematically. A generally accepted risk management is explicitly not required here[1]. No formal risk management structures or detailed documented processes are required at this point. The minimum requirement is to identify risks in due time and take targeted measures to keep them under control or eliminate them wherever possible. Elements of risk orientation must be found in all parts of the company and must be supported by documents and records.

In terms of content, this subchapter focuses on external risks (market development, innovations, interested parties) as well as on those risks that (can) influence the operation internally. The latter include process risks, planning risks, risks associated with customer or supplier relationships, resources as well as general risks associated with products and services that can occur in all phases of the life cycle beyond current orders and projects and after delivery.

---

[1]cf. EN 9100:2018, Annex A.4, p. 44.

© Springer-Verlag GmbH Germany, part of Springer Nature 2020
M. Hinsch, *Guideline for EN 9100:2018*,
https://doi.org/10.1007/978-3-662-61367-2_6

In addition to the risks, operational opportunities must also be determined in accordance with the requirements of Subsect. 6.1. The focus should be on a structured identification and evaluation of opportunities as well as follow-up actions according to the PDCA approach.

During the certification audit it must be made clear that the risks and opportunities are identified and actively and appropriately addressed using countermeasures. Organisations should therefore maintain records showing the identification, evaluation, measures, dates, responsibilities and past activities of risk management. Suitable objective evidence are for instance plans, evaluations, analyses and calculations.

Risks and associated countermeasures must also be addressed in the context of the management review (Sect. 9.3). For this purpose, an FMEA is most suitable, it may be supplemented by a risk matrix according to Fig. 6.1. A risk assessment should above all consider general company risks, process risks, market risks and personnel risks.

Since risks must be consciously identified, evaluated, mitigated and monitored within the framework of the risk-based approach, the next logical step leads to a systematic risk management. Companies should therefore consider whether they should (voluntarily) set up a simple risk management system in order to systematize the risk-based approach. This often facilitates risk control because *a single* company-wide process for operational and QM risks as well as for order and project risks (in accordance with Sect. 8.1.1) is employed.

6.2   Quality Objectives and Planning to Achieve Them

The quality objectives are derived from the link between politics and strategy on one hand and the provision of services in everyday business on the other. Quality objectives thus support the implementation of the quality policy at the operational level. Quality objectives must therefore be defined for the core processes, as well as for important support processes, departments or functions. They must not be determined half-heartedly once but be monitored at appropriate intervals and managed systematically.

An essential requirement of EN 9100 is that the objectives are measurable, thereby it is possible to clearly determine one's own quality status at any time. Moreover, progress in product and process quality can be detected over time. In addition to measurability, it is important that objectives directly or indirectly refer to process performance, product or service conformity or customer satisfaction. More generally speaking, the quality objectives should be focused on three factors: QM system, customer requirements and market needs.[2]

---

[2]See ISO 9001 (2016a), p. 1.

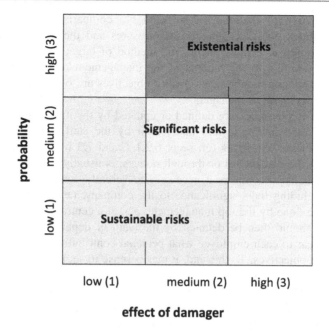

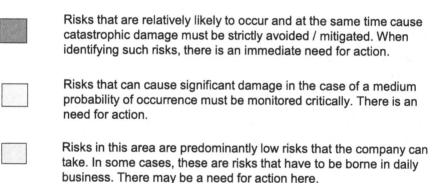

Risks that are relatively likely to occur and at the same time cause catastrophic damage must be strictly avoided / mitigated. When identifying such risks, there is an immediate need for action.

Risks that can cause significant damage in the case of a medium probability of occurrence must be monitored critically. There is an need for action.

Risks in this area are predominantly low risks that the company can take. In some cases, these are risks that have to be borne in daily business. There may be a need for action here.

Fig. 6.1 Simple risk matrix structure

The necessary activities for managing objectives are listed in Sect. 6.2.2 a)–e). Clear responsibilities, deadlines, resources and the way in which results are to be measured must therefore be specified.

An integral part of the definition and tracking of the quality objectives, is the at least annually conducted evaluation during the management review. If, however, the achievement of the objectives is to be monitored sustainably and effectively, a more frequent review is necessary. Companies with efficient QM systems monitor their KPIs against the objectives monthly, max. quarterly. If the KPIs or quality targets are only monitored annually, their desired control character is largely lost in accordance with the PDCA cycle.

The evaluation of the objectives includes a comparison of the target and the actual figures, the determination of measures and the determination of the resources required to achieve them. The method of target documentation is not specified. In addition to the records of the management review, (Excel)tools or ERP-systems are commonly used. Quality objectives are typically linked to variable salary agreements of managers.

If the quality objectives are defined or updated by the management, they must be communicated internally and understood by the staff. Employees must be made aware of the objectives (cf. sects. 6.2.1 f. and 7.3 b). Announcement, e.g. by e-mail, on the blackboard or through notices, is usually not enough. Creating awareness usually also makes it necessary to verbally communicate the quality objectives including their significance to the company, i.e. face-to-face. Ideally, this should be done by the top management, e.g. at a central staff meeting. These notifications should then be detailed by the team or department leader in order to make it clear to each employee what personal contribution he or she makes to achieving the objectives. To this end, it makes sense to set up a target system and to break down the targets from the management to the shop floor.[3]

In operational practice, it is often difficult for the person responsible (usually the QMR) to define suitable quality objectives, this applies especially to small and medium-sized enterprises. Above all, there is uncertainty regarding their EN suitability.

The following listing shows indicators that are adapted to the quality policy of our example in Subsect. 5.1 (*Quality policy and declaration of commitment*). These KPIs meet the EN 9100 requirements and can also be used to measure core process performance:

1. Sales hit rate outside Europe (ratio of quotations to incoming orders)
2. Quality of design planning (based on planned resources vs. those actually used => time, budget, man-hours)
3. Final acceptance rate in production and On-Time Delivery (OTD)
4. Supplier performance (goods receipt findings, OTD)

For example, the following measurable objectives can be derived from this:

1. The hit rate, i.e. the ratio of bids submitted to transactions concluded, is expected to rise from 35% to 38% in the following financial year.
2. The actual figures in terms of man-hours, completion date and budget for the largest three design projects are expected to be within a corridor of max. 95–105% of the planning data.
3. The nonconformities/error rate in the final product checks are to be reduced from 1.8–1.6%, OTD is to increase from 97.5–98%.
4. The rate of incoming goods findings is to be reduced from the current 3.2% to below 2.5%. The suppliers' OTD rate is to rise from 96–97.5%.

---

[3]See ISO 9001 (2016b), p. 2.

In addition to the product/service portfolio, the type and scope of the objectives depend primarily on the size of the company. For a production company with 50–70 employees, 7–10 objectives can be enough, which should then be broken down to the lowest working level. In contrast, a group with several thousand employees will normally require a target system across the various hierarchy levels.

Further information on the definition and measurement of process performance is given in the excursion of Subsect. 4.4 of this book.

When defining quality objectives for the first time, a re-definition of individual goals may sometimes become necessary in the first year of certification, because the quality objectives may be difficult to determine and measure or they may not be accepted. However, once defined, the objectives should not be altered if possible. Only the respective target value of the KPIs, must be improved continuously. The reason is that a complete change of objectives makes it difficult to compare quality progress over a long period.

In the certification audit, the auditor will ask about the measures and ways of achieving the objectives. After all, a definition of objectives alone does not mean the company takes process measurements seriously. The management must therefore be able to give a convincing answer to this question. During past audits, companies often saved their neck with very general statements. Here verifiable action plans and implementation examples are necessary. This applies in particular if KPIs were not achieved. Excuses will be of little help in the certification audit because the auditor must describe the achievement of the objectives in his audit documentation via the PEAR-form.

6.3 Planning of Changes

A QM system is not a static structure that is set up once and not changed afterwards. This is all the more true, since not only those changes that usually fall within the direct responsibility of the QM department that must be taken into account here. Elements of the QM system can be found throughout the entire organisation, including in value-adding processes. For example, changes to the ERP system or changes in procedures for the acceptance of products also affect the QM system. Therefore, planning in the sense of Subsect. 6.3 is necessary whenever changes are made that affect the conformity of products and services or customer satisfaction.

In these cases, it must be ensured that the changes are planned and implemented in a structured manner, considering the available resources and possible effects before implementation. Above all, this requires

- the evaluation of the type and scope of the change to the QM system,
- the influence of the change on operations and the potential effect on the conformity of products and services,

- measures/activities derived (including review and, if necessary, adjustment of tasks and responsibilities),
- the definition of responsibility and authorisation, and
- that the planning is adequately documented for the purpose of verification.

As a result, the efficiency of the QM system must not lose effectiveness. All too often the QM system is suspended for the duration of the implementation period for the sake of simplicity. This is not permitted.

References

CEN-CENELEC Management Centre: EN 9100:2018—Quality management systems—Requirements for aviation, space and defence organisations. Brussels (2018)

CEN-CENELEC Management Centre: EN 9120:2018—Quality Management Systems—Requirements for aviation, space and defence distributors. Brussels (2018)

International Organisation for Standardisation and International Accreditation Forum: ISO 9001 Auditing Practice Group: Guidance on: Policy, Objectives and Management Review. 13.01.2016 (2016a)

International Organisation for Standardisation and International Accreditation Forum: ISO 9001 Auditing Practice Group: Guidance on: Improvement. 13.01.2016 (2016b)

Support

7

Chapter 7 deals with supporting elements for service provision: These include human resources in terms of quantity and qualification (Subsect. 7.2), the infrastructure and the working environment. Awareness (Subsect. 7.3) and operational knowledge (Sect. 7.1.6) are becoming increasingly important. Finally, Chap. 7 defines documentation requirements (Subsect. 7.5).

Chapter 7 focuses on the basic and long-term provision of resources. Requirements for short-term, order-related resource availability are formulated in Chap. 8.

7.1   Resources

7.1.1   General

The management must ensure that the human, infrastructural and financial resources required to introduce and maintain a QM system in accordance with EN 9100 are made available on time.

The long-term determination and procurement of resources is in the responsibility of the management within the framework of the predominantly multi-year organisational planning.[1] The short-term determination and provision of resources, on the other hand, is carried out by planning at an operational management level. Based on the current order situation, the responsible departments organise the availability of personnel, machines, equipment and resources, storage areas and

---

[1]Even if this long-term planning (budget or annual planning) ensures a periodically recurring determination of requirements, the provision of resources should also be addressed and documented in the management review (e.g. capacity adjustments or important new acquisitions, see Sect. 9.3.2 b).

© Springer-Verlag GmbH Germany, part of Springer Nature 2020
M. Hinsch, *Guideline for EN 9100:2018*,
https://doi.org/10.1007/978-3-662-61367-2_7

suppliers. A company may maintain a permanently efficient QM system only through short- and long-term planning activities as well as their implementation.

Certification audits sometimes show that insufficient resources are available during the year to maintain the QM system in accordance with EN 9100. Particularly regarding the realistic evaluation of the resources required to support the QM system during the year, some top managers assume that the costs for the certification auditor or certification body are high enough and that further expenditure would neither be necessary nor reasonable. However, the EN 9100 requires capacity between the certification audits, in particular for target tracking and process control as well as for the fulfilment of the comprehensive documentation requirements, e.g. in the context of corrective actions, for the handling of non-conformities, the evaluation and monitoring of suppliers, the management review, risk control or the planning and monitoring of personnel qualification. Deeper root-cause analyses are also a time-consuming element of an efficient QM system and are all too often neglected in operational practice.

Section 7.1.1 of the standard is formulated in very general terms and contains no concrete instructions for action. For those companies that manage their resources systematically and considering all important internal and external information, the very broad formulated standard requirements of Sect. 7.1.1 can be ticked off as fulfilled without any need for further action.

Examples of Inadequate Resource Allocation
Whether the requirements of Sect. 7.1.1 are fulfilled cannot always be deduced from a single observation, but often provides an overall operational picture.

Example 1: During the audit, the QMR states that, in addition to his operational tasks as a materials planner, he has too little time for the concerns of quality management. At the same time, the QM system is in an unsatisfactory condition because no internal audits have been conducted since the last certification audit, a process measurement does virtually not exist.

Example 2: A company avoids purchasing an ERP-system and instead works with post-its and complex Excel tables. This leads repeatedly to incorrect or unpunctual deliveries. The infrastructure is therefore not in an adequate condition. The managing director is of the opinion that modern IT only costs money and that the status quo has always worked well.

7.1.2 People

An important prerequisite for ensuring high product and service quality is sufficient personnel availability (quantity) on one hand and appropriate personnel competence and qualification (quality) on the other.

An optimal personnel *capacity* results from the operational planning and the workload. The necessary personnel *quality* is derived from the type of activities to be carried out. Well-trained employees are not only necessary from the EN 9100 point of view, but also from an economic perspective. After all, qualified personnel generate a higher work output at a given period while reducing the number of errors made, so that the cost of quality issues due to improper work performance decreases.

Furthermore, in the event of an accident, appropriate qualification of the personnel can facilitate the release of managers from liability under labour and civil law, because the fulfilment of organisational and supervisory duties has been ensured.

Section 7.1.2 of the standard is formulated in very general terms and contains no concrete instructions for action. The proof of an adequate personnel planning regarding the quantity can be shown by planning tools or Excel files.

Further requirements regarding staff qualification, i.e. competence and awareness, are contained in Sects. 7.2 and 7.3 of the standard and can therefore be neglected here.

7.1.3 Infrastructure

Companies with an EN 9100 certification must have facilities with equipment conforming to the size, type and scope of service provision. The infrastructure includes:

(a) Offices, workshops, test stands and hangars as well as workplaces and storage areas, warehouses, bathrooms, kitchens and rest areas, incl. heating and ventilation systems as well as energy and water supply,
(b) Resources such as machines, measurement instruments, tools and equipment, storage systems, office facilities, safety and rescue equipment,[2]
(c) Transport systems and transport structures for inbound and outbound deliveries,
(d) Communication systems such as telephones and e-mail as well as
(e) IT infrastructure (hardware and software) including data backup systems and appropriate data security.

The infrastructure shall be regularly checked for adequacy and condition as well as for its completeness. Responsibility for monitoring and evaluating the infrastructure lies with the management and is therefore (an indirect) part of the management review.

In parallel, infrastructure with high operational relevance should be identified and included in the risk analysis (chap. 6) in order to assess the consequences of

---

[2]E.g. fire extinguishers, first aid kits, eye wash bottles.

non-availability. Emergency response measures must then be available in the case of the failure of important machines. This may involve, for example, keeping a stock of spare parts, keeping the options to choose between suppliers on a short-term or ensuring the short-term availability of service technicians from equipment manufacturers.

It is not necessary for the infrastructure to be owned by the organisation. Nevertheless, a demand-oriented availability is important, which can also be ensured by means of leasing or short-term renting.

Appropriate infrastructure may not only be procured and used for product, service and process requirements, but it may also be used for safety, economy, reliability and maintenance.[3] For equipment requiring regular maintenance, availability of maintenance schedules and records must be ensured.[4]

The IT infrastructure is sometimes identified as a separate item on the audit agenda, in most cases it is at least separated from the rest of the infrastructure. Nevertheless, IT is often not given the necessary attention in certification audits. This should not prevent companies from focusing on the IT infrastructure carefully; after all, operational assets are usually not primarily tied up in tangible assets but stored in the form of data in the IT system. Therefore, it is not only important with regards to the certification audit that the company has clear rules for the protection of the operational data.

The following should be defined in writing: Requirements for password selection, unauthorised software installations, access rights, procedures for destroying or decommissioning data media (hard disks, DVDs, etc.) as well as general information (or training) to sensitize employees to risks. To this end, it is particularly important to create an awareness of the risks and weak points of an IT infrastructure in general. Therefore, it is for example advisable to train the risk from the www and from emails, paying special attention to their attachments.[5] Much of what needs to be trained may sound natural to some, however, staff knowledge in terms of IT security requirements is often frightening.

Beyond employee qualification and awareness, security software (firewalls, virus protection, etc.) must be set up in an extend that is appropriate to the value of the data to be protected.

---

[3]See ISO 9004 Subsect. 6.5.

[4]Many modern machines are self-maintaining. Here it is sufficient to document only those maintenance and repair measures which are carried out beyond the mechanical self-maintenance (see also Sect. 8.5.1.1).

[5]For example, the opening of zip or exe files as well as MS Office files with macros from emails should generally be prohibited - even if the sender is supposedly safe. The author can report about a customer with whom a trojan had nested because a purchasing employee opened the zip attachment of an actual supplier that was titled "invoice". The supplier's network was hacked, the email externally controlled. The trojan was found very quickly in the present case. Nevertheless, all systems had to be shut down for 36 h to check the distribution rate of the trojan. During this time the production stopped.

Due to complex, often undetected IT attacks, extended measures such as the encryption of network connections may be necessary. In addition, emergency management is recommended to determine how to react on loss of sensitive data or in the event of hacker attacks.

7.1.4   Environment for the Operation of Processes

The provision of services must take place under "controlled" environmental conditions in order to guarantee appropriate product or service quality. This means, first of all, that the working environment must not cause any restrictions in process performance, excessive distraction of personnel or negative impacts on the use of resources. The working environment mainly includes the following areas:

- order and cleanliness (see also 5S–Sort, Set in Order, Shine, Standardise, Self-discipline),
- appropriate temperatures, humidity, ventilation,
- year-round protection against weather conditions (wind, rain, snow, ice, sand),
- as little dust and other air pollution as possible,
- adequate lighting,
- minimum, but at least justifiable noise level,
- workplace-specific precautions regarding product preservation (e.g. ESD precautions),
- compliance with occupational safety regulations and health protection requirements as well as workplace-specific precautions regarding environmental protection (e.g. in non-destructive testing workshops, suction devices when working with hazardous substances, protective goggles when operating grinding machine).

However, the working environment not only includes the physical environment, but also social and human factors. Employees should be brought to optimal performance through an appropriate and motivating working environment. The standard specifies the following conditions:

- creation of organisational structures that promote communication and teamwork,
- avoid a lack of attention due to fatigue & exhaustion,
- ensuring motivational working conditions,
- employee-friendly handling of social norms,
- minimise pressure, stress and distraction.

7.1.5   Monitoring and Measurement Resources

In order to ensure that products and services meet the defined requirements, it is necessary to conduct monitoring and measurement activities. Therefore, appropriate resources need to be available. That can be:

1. monitoring and measuring equipment (also: test equipment),
2. documents (e.g. checklists, samples) as well as
3. qualified personnel.

This section focuses on the "typical" monitoring and measuring equipment. However, documented information may also fall under this heading and be evaluated for this purpose. For example, a photograph serving as a reference shall be validated. However, using documentation as monitoring or measurement resources, strongly overlaps with Sect. 7.5.2 c) in the context of preparation and updating of documented information. The requirements of Subsect. 7.2 on personnel competence apply to the qualification of test personnel.

Monitoring and measuring equipment must be checked at regular intervals for condition and accuracy in order to ensure that it is fully functional in the long term. This requires the implementation of a process for sound management and systematic monitoring of measurement equipment.

Introduction of Test Equipment
The correct management of monitoring and measuring equipment begins with the receipt of goods. Thus, the inspection equipment must be correctly registered after the goods receipt inspection, e.g. by an employee in the tool issue department or quality management. For this purpose, each piece of test equipment has to be appropriately marked (inventory number and, if applicable, department code). Test equipment requiring calibration should additionally be labelled with a sticker showing the expiry date of the calibration.[6] In accordance with Sect. 7.1.5.2 c) of the standard, monitoring and measuring equipment shall be secured against unauthorised modifications wherever possible. In addition, the tools are to be protected adequately during periods of non-use.

In addition, monitoring and measuring equipment must be listed in a register (equipment database) before it is used for the first time in order to ensure its tracking. The following shall be recorded:

- type of equipment,
- inventory number for unique identification,
- place of storage or use,
- monitoring interval[7],

---

[6]Not all test equipment that can be calibrated is used for quality tests on the customer's product, e.g. that used in pre-development or for training purposes only. For this test equipment, a lower accuracy may suffice, so that calibrations are not necessary. In order to avoid any risk of confusion with measuring instruments for quality tests, such equipment shall be labelled with a clearly visible sticker indicating that the measuring instrument may not be used for quality tests.

[7]Type and frequency can be found in the equipment documentation (e.g. manufacturer's manual). In individual cases, specifications can also be defined by the customer or the aviation supervisory authority.

- acceptance criteria and,
- information on any associated documentation (e.g. equipment manual, special calibration requirements).

Employees must always have access to the corresponding operating instructions and equipment manuals for their day-to-day work.

Before using complex monitoring and measuring equipment (e.g. oscilloscopes) for the first time, the employees concerned should be instructed. Training is the best way to ensure that the equipment is used carefully and in accordance with regulations.

Monitoring and Testing Test Equipment
The monitoring and measuring equipment shall be continuously monitored via the equipment register or database. A device must be retracted at the end of the defined test or calibration interval in order to determine whether its functionality and accuracy still meet the requirements. Expired measuring equipment may no longer be used. The periodic recall for device testing and calibration must be clearly defined as a process.

Corresponding (technical) test instructions shall be available if calibrations are performed by the organisation itself. For this purpose, the specifications of the test equipment manufacturer can be used. In most cases, test equipment manuals describe measuring conditions, intervals, test procedures as well as tolerances or acceptance criteria. When performing equipment verification or calibration, it is to be ensured that the test meets the specifications and that calibrations are performed according to an officially recognised standard/measuring standard.[8] If these are not available, the basis for the test or calibration must be documented ("Against what was the test performed?"). This proof must be archived in order to make the evaluation traceable. After testing or calibrating, the calibration status on the tool should be updated, even if traceability via the equipment list is enough formally. A round sticker with the month and year of the next inspection is usually used for this purpose.

In many cases, tests, calibrations and maintenance of measuring equipment are also carried out by the manufacturer or external specialists.[9] In this case, the organisation simply must ensure that the test equipment is collected on time and sent to the specialised company.

For the purpose of traceability, records shall be kept of the tests and calibrations performed (minimum test record or calibration confirmation). After completion of the internal or external test activities, the information in the equipment database must also be updated.

---

[8]Note: This is a frequently evaluated criteria in certification audits.

[9]These companies must be qualified for the corresponding calibration. When selecting suppliers, care should be taken to ensure that the subcontractor is appropriately qualified, e.g. by certification according to ISO/IEC 17025. However, the use of an accredited calibration laboratory is not explicitly prescribed in the EN 9100.

The archiving period must be longer than the inspection interval and should be at least three years.

Inadequate Test Equipment

If a monitoring or measuring device no longer meets the requirements, it must be repaired or, if necessary, permanently withdrawn. In this case, it must not be forgotten to delete the device from the equipment database and to document the deletion.

In the case of malfunctioning measuring equipment, it must be checked whether the restrictions in functional capability had an influence on the test results of previously tested products. This is also necessary if systematic inspection errors have been identified resulting from incorrect specifications (e.g. incorrect tolerance data) or insufficiently qualified personnel. For this purpose, the focus should first be directed towards records of earlier measurements in order to limit the temporal scope of the deficiency. Traceability is ensured if the equipment number of the measuring device used is documented on the job card or work order. Therefore, the organisation must have a procedure that defines the handling of nonconforming measuring equipment and, above all, the products concerned. For example, the extent to which measurements must be repeated or a recall of products already delivered must be determined. In practice, many companies lack this aspect of the measuring equipment process.

Example: Defective Torque Wrench

During the calibration of a torque wrench for an engine component, it was determined that the torque wrench was no longer correctly calibrated. As a result, all nuts had been tightened with a significantly increased torque. A crisis team was convened, and all affected engines were identified (the tool number was recorded in the job cards). The risk assessment revealed that the risk could only be fully assessed with the support of the OEM. The engine manufacturer found that in the worst case the nuts could have been broken and the engine could have failed catastrophically. Due to the criticality, the responsible aviation authority was then informed. The authority finally issued an Airworthiness Directive (AD), which required the replacement of the nuts for the affected engines within a short time.

7.1.6   Knowledge of the Organisation

In companies, knowledge is at least as important as machines, plants and equipment. Knowledge is power and is equated with business success. An awareness of the *existing* operational know-how (*actual*) on one hand and the *required* knowledge (*objective*) on the other hand forms the indispensable basis for this. This is all the more true for service industries. Therefore, every company should consider the following questions:

- What knowledge is required for the provision of services or its processes?
- How is organisational knowledge secured?
- What are the sources for updating knowledge and how is new knowledge implemented in the organisation and finally integrated into the products or services?
- How is knowledge lost and how can it be protected?
- What advantage does the company have over customers and competitors; where does the company depend on the know-how of others?

These are the questions every company is confronted with, regardless of its range of activities and size. Nevertheless, knowledge is taken for granted in many companies without paying attention to it. Management must therefore implement systematic structures to ensure that operational knowledge is built up, maintained and secured.

Knowledge management does not have a single measure. Knowledge transfer is ensured with tools and measures such as internal knowledge databases, wiki-systems, mentoring programs, conscious knowledge transfer communication or the recording and implementation of lessons learned findings. In contrast, fluctuation of personnel and the inability to secure or share knowledge are risk factors and should be treated as such.

Knowledge of processes, products and services can be maintained by means of process descriptions, work instructions as well as product or service specifications or wiki systems. A sales or trade fair, projects with universities or joint ventures, company or patent acquisitions are examples of how knowledge can be expanded. Protection against the loss of knowledge can be provided, for example, by means of IT security, individual access authorisations or intellectual property (IP) management. Another option is to conduct strategic personnel planning for 10–20 years into the future in order to anticipate knowledge losses due to retirement or employee fluctuation in a timely manner and thus keep it under control.

To achieve this, the necessary know-how must be systematically identified, communicated, preserved, expanded, updated and protected.

The need for such planning is often evident in business practice when companies take former employees back from retirement because the knowledge was not handed over in time to the younger generation. Such a development indicates risks due to inadequate planning.

7.2 Competence

Systematic personnel competence is an essential prerequisite for high product quality and safety. Only well-trained employees can ensure that operational processes run smoothly over a long period of time while continuously improving.

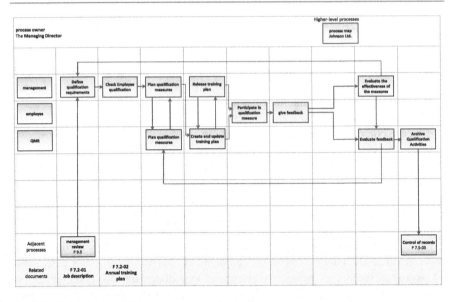

Fig. 7.1 Exemplary qualification process

In this respect, every company is ought to develop concepts of how to ensure an appropriate qualification of its personnel. A structure and a procedure must be defined for this. A rough concept or methodological framework, e.g. in the form of guidelines and templates (qualification profiles, action plans or induction plans), should exist for this purpose. This can be documented in the QM manual or in a qualification process (cf. Fig. 7.1). It must show how initial qualification and initial and continuing training are ensured on an ongoing basis.

The concept (cf. Fig. 7.2) should be appropriate to the size of the organisation. In daily practice, this means that often only medium-sized and larger companies maintain a qualification system with a structured concept. In SMEs, however, the qualification concept is often defined in a short process description. The effectiveness of the framework shall be demonstrated by appropriate training records.

In the certification audit, it is important that the managers interviewed can demonstrate the competence of their personnel. It must be demonstrated that the qualification process is capable of empowering staff to fulfil their tasks. The key is to qualify employees for what they are actually doing and not for what they should be doing. Finally, the personnel qualification process must ensure that changes in the responsibility are identified and, if necessary, qualification measures are derived.

(a) *Determination of personnel competence*

The determination of personnel competence is like a medal with two sides. On one hand qualification requirements must be defined as the target requirement of a

Fig. 7.2 Exemplary basic structure for a detailed qualification concept

job. For this purpose, it must be determined what is to be done in the job in question. On the side, it must be defined which skills are required of an employee in order to be sufficiently competent for the job or activity concerned. On the other hand, the knowledge and skills of the employees concerned have to be determined in order to get an overview of the actual competence. Both tasks are normally the responsibility of the supervisor, and if necessary, in cooperation with the personnel department.

The more complex part of the competence determination is the definition of the target requirements. For this purpose, a job description with the qualification requirements should be derived. This is the basis for the definition of competences, responsibilities and authorisations. At the same time, a job description helps to structure and harmonize job requirements in terms of knowledge and skills.[10] Job descriptions thus form an important cornerstone for meeting standard requirement 7.2 a).

The difference between actual and targeted competence indicates the additionally required qualification of an employee for the position in question. In order to close the gap, qualification and induction plans are necessary instruments that must be demonstrated in a certification audit. The plans shall include information on

---

[10]Job descriptions should be signed by the job holder when taking up the position. Although this is not required by the standard, it is an appropriate proof that the employee is aware (or should be aware) of his or her qualification requirements, the area of responsibility, and the scope of authorisation. The signed job description must then be archived in the personnel file.

- necessary on-the-job training,
- job-specific training, instructions and further education (e.g. in the operational QM system, software training, machine operation or occupational safety) as well as
- the planned or actual date or period of implementation.

The determination of the exact qualification requirement and period is usually the responsibility of the line manager with the support of the employee concerned.

Example: Qualification Requirements for a Work Planner
- Successfully completed technical vocational training as an aircraft mechanic or electronics technician (m/f) or comparable qualification
- Several years of relevant professional experience as well as sound experience in planning processes and project management
- Knowledge of SAP, MS Office, AutoCAD
- Strong in communication and assertiveness, team player, very good judgement regarding technical and business contexts
- Fluent in spoken and written national language and English

In addition to initial training, it must also be clear whether periodic re-training or refresher courses may be necessary after completion of the qualification. If retraining is required, the extent also has to be clarified. Later qualifications may become necessary due to the acquisition of new machinery or equipment through new or changed processes and procedures or due to a change of responsibilities. It may also be necessary to extent the qualification for experienced employees, for example in case of

- a new customer with new requirements,
- expanding the range of services provided by the company or
- changes in legislation.

In this respect, a (brief) review of personnel competence is to be carried out not only for new appointments, but also on a regular basis (see also the NOTE in Subsect. 7.2). Such examinations can take place, for example, in the course of staff appraisals, during the acquisition of new orders or during the planning of annual training.

The standard focuses not only on the company's own permanent staff, but also on all employees who carry out activities under the supervision of the company. In this respect, temporary workers may also be included. Special attention is also required because temporary employees often do not have the same familiarity with operational procedures as the core workforce. It must also be considered that the employment agencies for temporary workers may not always be able to provide personnel with the qualifications originally promised. Therefore, it may not be sufficient to rely solely on the promises made by those agencies. Appropriate proof of qualification must be requested and checked. If the documentation is unclear, separate qualification tests must be carried out and their results recorded.

Adequate competence of external personnel is not only meaningful because these personnel often influence the quality of products and services but also for liability reasons of the executives regarding their supervisory duty

(b) *Ensuring an appropriate qualification of personnel*

A closer examination of this requirement of the standard can be omitted. It is not necessary. If the employees are sufficiently competent, this EN subchapter is fulfilled and if the staff needs to be qualified, requirement 7.2(c) shall apply.

(c) *Implementation and evaluation of qualification measures*

Employees do not always meet the full range of operational and job-specific requirements but must first be introduced to their tasks. This is done by imparting theoretical knowledge as well as practical skills and experience. In doing so, the personnel learn both the technical and the non-technical skills, i.e. the interpersonal requirements of the respective job.

The type and scope of the employee qualification must be such that the personnel are enabled to carry out the assigned tasks independently and in a quality that meets the requirements. There is no universal definition of appropriate employee qualification. This depends on the respective job and the previous experience, the knowledge and skills as well as the individual's perception and understanding.

However, a certification auditor will generally assume that the qualification is inadequate if there is an accumulation of execution errors in the same activity or by the same employee. Sample evaluations of the workplaces or the documentation also repeatedly lead to audit findings because personnel are not sufficiently qualified for the tasks performed by them (e.g. lack of training plan, forklift driving license or machine/equipment instruction cannot be proven).

Training and Qualification Planning

If the competence level does not correspond to the level required by the job, the organisation must ensure that corrective measures are taken. Education and training may include the following elements:

- (theoretical) basic training (e.g. specialist training or courses),
- on-the-job training (practical experience),
- supplementary qualification measures (these include, for example, training or instruction on operational procedures and equipment),
- recurring/continuation training (e.g. human factors).

The overall operational training needs shall be summarised in a training plan. The purpose of this is to control the qualifications activities and to enable the timely provision of financial resources.

In most companies, the training plan is created at the end of the year along with the budget planning and is updated during the year. Updates during the year not only show newly added qualification measures, but also show whether and when planned events have already been carried out.

In many cases, small and medium-sized enterprises (SMEs) find it difficult to plan annual trainings because only a few trainings are held at short notice when required. For SMEs an annual planning horizon is too large for training plan. In practice, therefore, training planning usually plays a minor role. In the certification audit, a simple planning or overview with the few known training measures together with an overview of the trainings and instructions already carried out during the year is enough.

Effectiveness Control

After the qualification measure has been carried out, an effectiveness control must be conducted. Sometime after the measure has been implemented, the superior must check whether the training contents have been understood by the employee and whether they are being applied by him or her in everyday business life. The effectiveness control must be documented. For this purpose, a note in the training planning or training records or in the documentation of annual staff appraisals are suitable. A note shall be made on the results of the effectiveness control with date, inspector and test object (e.g. order number). If the rating is poor, retraining must be ensured.

Certification audits often show that the effectiveness of qualification measures is either not assessed at all or is not documented. There must be an effective PDCA cycle in place, which is convincing throughout all stages.

(d) *Documentation and archiving of qualification measures*

In order to provide evidence of an appropriate personnel qualification, records must be kept of the knowledge and experience of the employees. This applies both to the employee's previous qualification in the job and to training measures, instructions and practical experience gained after taking up the current position. Suitable supporting documents are certificates, conformations and references. Internal qualification measures are often tracked by signed participant lists and logbooks or training records for practical proof of qualification over a longer period. To ensure that operational management also has a structured overview of personnel qualification, a qualification matrix should be maintained in large departments with similar tasks (e.g. in production or engineering).

Aviation legislation can used as a guideline for the archiving period, according to which a retention period of two years beyond the duration of the employment is prescribed for personnel data.[11]

---

[11]EASA (2003), AMC 21A.145(d) (2) and EASA (2003), AMC 145.A.35 (j).

7.3 Awareness

Since its 2018 revision EN 9100 pays greater attention to employee awareness through the introduction of a separate subchapter. The focus is on improving quality awareness. The aim is that the staff gains a better understanding of their own actions. For example, employees must be able to assess when products or services meet the requirements and moreover, they must be aware of the consequences of nonconformity. In addition, an appropriate QM awareness requires the familiarity of the employees with the characteristics and core elements of the QM system, such as

- customer orientation,
- process orientation, and
- risk-oriented action.

For this it is important that the QM system including its objectives, its processes, procedures, aids and specifications are not only available, but are also understood by the employees in type and scope. By doing so companies can succeed in anchoring a sustainable awareness of the importance of the elements of a functioning QM system in the minds of their employees.

A company-wide awareness down to the employee level is most likely to be achieved if strict quality orientation becomes part of the organisational culture. Additionally, to incorporating quality aspects in organisational training, it is necessary that quality orientation finds its way into everyday business life. This is only possible with an unambiguous and recurring commitment to quality on management and executive level! With the commitment of the management, success stands and falls for creating awareness among the employees. Direct communication between management and employees is essential for this. Campaigns with info letters, posters, flyers or videos are only a second-best solution for ensuring an awareness, as they remain ineffective if the management shows no credible commitment.

As a specific measure to create a comprehensive quality awareness, the standard requires appropriate awareness for quality policy and quality objectives (Subsect. 7.3 a and b)—both from internal employees and external personnel under company supervision. The standard thus formally requires an understanding of quality policy and objectives by anyone outside the organisation who works under the supervision of the company—irrespective of the type and duration of the employment.

The extent to which the requirements of Subsect. 7.3 are sufficiently addressed can be quickly checked by a certification auditor, as compliance is by no means merely the portrayal of an overall picture. Quite a few auditors randomly ask employees about quality policy and objectives. "Never heard of this" is the wrong answer and will lead to an audit finding if the response is given repeatedly. Employees must be familiar with the general elements of the quality policy and in detail with the quality objectives relevant to them. However, since a lot of employees have problems naming content and storage in many organisations, quality policy and objectives should be displayed on the notice board or, better still, in exposed places where they can be read by everyone. Quality policy and objectives are accessible to every employee at these locations.

The certification auditor can also easily check an appropriate awareness for service provision, the functioning of the QM system (c, d, e) as well as product safety (g). For example, employees must be aware of the basic requirements of quality management. In addition, employees must be aware of their own role and contribution in achieving high quality, product conformity and product safety. Furthermore, they must be aware of the consequences of their own mistakes for the customer (f).

In addition to the requirements for an awareness of typical QM aspects, g) and h) of the standards Subsect. 7.3 contain aspects on compliance and ethical conduct. However, it is not so much social-ethical behavior that is demanded as it is product- or company-oriented ethics. The IAQG gives the following examples:[12]

- Reporting and active problem solving instead of waiting out errors and nonconformities,
- Preventing a culture that tolerates unethical behavior even in the slightest, as this favors all kinds of illegal and even criminal activities,
- Respecting and complying with laws, government regulations, standards and internal rules. Conflicts of interest, accepting gifts, invitations or other favors from customers or suppliers, respecting export restrictions, compliance with intellectual property agreements are dealt with,
- Introducing an organisational culture that conveys the organisational roles to the employees.
- Employees should therefore become aware of what the company represents beyond its economic and technical requirements. This also includes communicating the values that define the company. These values are often already reflected in the quality policy. These include, for example, employee appreciation, equal rights or sustainability. But also, excluding the use of controversial minerals or military orders fall may under the ethical requirements.

7.4   Communication

It is the responsibility of the management to ensure appropriate communication within the company and with external parties. Ways of appropriate communication are personal conversations, email exchange, exchange of information via the intranet, telephone, company letters, info sheets or notices. Appropriate

---

[12]See IAQG (2016a), p. 12.

communication must not only function vertically in the hierarchy but must also be ensured horizontally between departments and teams.

Ensuring adequate communication appears to be obvious at first, but deficits are regularly identified in everyday business. Especially in larger organisations as well as in autocratically managed companies or teams, communication is often only just adequate. Across all organisational sizes, shortcomings often appear in the communication of strategy issues and quality management. For example, employees often lack knowledge and awareness of quality in general and knowledge and understanding of quality policy and quality objectives in particular.

In this respect, communication structures and standards must be implemented that define how specific information is circulated by whom, through which channels, to whom and to what extent. There may be more than just *a single* communication process. The operational communication structures can be described in many processes (e.g. sales or planning processes). It is important that employees are aware of the requirements regarding the type and scope of communication in their respective areas of responsibility. However, beyond all formally defined communication channels, an open communication culture should be established, as it is the most effective means to avoid a lack of communication.

The standards' communication requirements apply internally as well as to external parties. It must therefore also be defined how, when and, above all, via whom communication takes place with customers, suppliers and interested parties (for customer communication, see also Sect. 8.2.1).

In the certification audit, the effectiveness of communication will not be audited as a separate process, but usually as an embedded element in the various processes and departments or teams. A *systematically* insufficient communication, from which a serious audit finding can be derived, may be determined but is difficult to prove. In order to prevent this from happening, protocols, meeting minutes or emails should be stored. Moreover, it is advisable to create a communication matrix in which at least the company meetings and internal events (e.g. works meetings) are stored along with frequency and participants.

7.5   Documented Information

7.5.1   General

Standards are as famous as they are notorious for their documentation requirements. In EN 9100 these can be found in all chapters, whereby the basis is defined in Subsect. 7.5.

Quality management distinguishes between documents and records. The former are specifications, the latter are compliance documents. Records are therefore also documents but are usually categorised separately.

The standard uses the superordinate term "documented information". This is to express the fact that the type of medium for documentation is irrelevant. For example, the quality policy or a work instruction can also be available as a video or audio file. However, documented information may be specified in the following categories:

- Operational QM documentation (e.g. process descriptions, work and process instructions, templates, filling instructions and empty checklists, job descriptions, videos),
- (internal) technical documents (e.g. own manufacturing or maintenance instructions, for example drawings, diagrams, test descriptions and specifications, sample photos, videos),
- external documentation (e.g. customer specifications, operating or maintenance instructions, design documents, drawings, circuit diagrams, standards, laws, regulations),
- Recording/proving documents (e.g. certificates, minutes, approval/acceptance documents, certificates of execution, video recordings, completed checklists).

In *Note* in Sect. 7.5.1, it is pointed out that the type and scope of the documented information are based on the individual circumstance of the company. The key factors mentioned there, are the size of the organisation, the product portfolio or range of services, the complexity of the service provision and the skills of the personnel. In certification audits, the scope of documentation can lead to discussions between companies and the auditor because there are often conflicting opinions as to the level of detail in which QM documentation needs to be available and records should be kept. This applies in particular where documented information is not explicitly required by the standard. As a rule of thumb, however, the phrase *No record, no evidence* always applies.

The reluctance towards recommendations or findings of the certification auditor should be limited here. The auditor can often provide a better assessment of what is acceptable than the industry standard. In addition, conflicts regarding the correct scope of documentation can adversely affect good audit atmosphere and ultimately lead to a more sensitive audit elsewhere. If there are different views, it is usually advisable to agree with the auditor and to minimise efforts when rectifying the finding. Thereby none of the involved parties lose face.

Specifications and Guidelines

According to Sect. 4.4.2, companies are obliged to define their processes in writing. Irrespective of the size of the company, there is a minimum requirement of 15 to 20 process descriptions which must be available even in the smallest organisation.

However, companies do not necessarily need a QM manual. This is no longer explicitly prescribed. However, due to the high documentation requirements of EN 9100, the provision of a QM manual is recommended. Finally, a QM manual also has its advantages: It provides employees and external auditors a solid overview of the operational quality management system and the most important quality requirements *one* reference document:

- Quality policy and quality objectives (Subsects. 5.2 and 6.2),
- Declaration of compliance by the top management (Sect. 5.1.1),
- Definition of the scope of application (Subsect. 4.3)
- Organisational structure with key responsibilities and authorisations (Subsect. 5.3),
- Implementation of the process-oriented approach specific to the company as well as a description of the processes relevant to the organisation, incl. a subordinate process-map (Subsect. 4.4),
- Profile of the organisation including an overview of operational resources to provide external workers or new employees with a brief overview of organisational activities and facilities.

In addition, the standard contains further indications where a written documentation is advisable. According to EN 9100 process have to be determined for:

- Metrological traceability (Sect. 7.1.5.2),
- Back-up process for electronic data/information management (Sect. 7.5.1)
- Planning of product and service realisation (Subsect. 8.1),
- Handling of critical items (Subsect. 8.1),
- Control of outsourced processes (Subsect. 8.1),
- Management of operational risks (Sect. 8.1.1),
- Configuration management (Sect. 8.1.2),
- Design management (Sect. 8.3.1),
- Process for supplier selection, evaluation and control (Sect. 8.4.1.1),
- Monitoring of externally created documents and records (Sect. 8.4.1.2),
- Evaluation of test reports for outsourced special processes (Sect. 8.4.2.1),
- Production or provision of services (Sect. 8.5.1b),
- Special processes (Sect. 8.5.1.2), and
- Handling nonconforming products (Sect. 8.7.1).

Formally there is no obligation for a written process. However, the EN 9100 requirements will be difficult to implement and maintain if there are no clear and comprehensible instructions for the employees. If written requirements are not explicitly prescribed, they facilitate the verification in the audit. After all, the principle *no record, no evidence* applies to the certification audit.

Compliance Documentation
Records are a special type of documentation. Records are compliance documents, that prove whether, how or with what results tasks, activities or work steps were carried out. They can also be used to provide information on the condition of products or services. Records include, for example, completed checklists or forms, certificates, documented measurement results or stamped work orders. EN 9100 contains numerous references to "documented information" which call for written records. In principle, direct and indirect compliance documents of any kind must exist for most chapters of the standard. Maxim: *no record, no evidence.*

7.5.2   Creating and Updating

This section of the standard lays down rules for the preparation and updating of documented information.

(a) Documents and records shall be adequately identifiable and described. For this purpose, minimum information is given as examples in the EN text. Other important information may include the revision status, the date of issue or any period of validity. Appropriate labelling is used to ensure the clear identification of documented information so that its history can be traced easily.

(b) It is irrelevant whether the company makes documents available in paper form, as pdf files, via the intranet in html format, as video, as audio files or in any other way. Likewise, the type of storage and archiving medium (paper, web, tapes, film) is not prescribed. Language may play a role in documents and records because employees need to understand what they are reading. If English is prescribed as the working and documentation language, it must also be ensured that all employees concerned understand and (can) apply it. It is important that the documented information meets the requirements of customers and legislators.

(c) All documented information must undergo a release process before it is officially published in the company. This ensures that only documents that are complete, accurate, necessary and evaluated by a competent and authorised person are released for circulation. In accordance with the *Note*, employees or bodies authorised to release documents shall be identified. In addition, before releasing a document, the following aspects must be considered in addition to the correctness of the content:
   - Consistency with other requirements,
   - Consideration of references,
   - Fulfilment of design and customer requirements,
   - Applicability or feasibility of specifications,
   - Compliance with legal, regulatory or normative requirements.

A written procedure must be defined for handling documents and records. This can reduce the risk that unchecked QM specifications or QM documents approved by unqualified or unauthorised employees, as well as technical or order-related documents, will be circulated.

7.5.3   Control of Documented Information

7.5.3.1 Availability and Protection of Documented Information
EN 9100:2018, Sect. 7.5.3.1 provides the following requirements for the handling of documents and records:

(a) It must be ensured that all documented information required for the work execution is available in proximity of the respective workplace. This is less about records than about documents, such as QM specifications but also technical or order-related documentation. Documented information must be stored in a sufficiently comprehensible manner, and it must of course be ensured that it can be found again. In operational practice, it sometimes happens that the QM documentation is password-protected (especially via the intranet) and thus not accessible for employees without an IT account. Another negative example is the storage of material/operating material data sheets by the responsible material planner and not on site by the users (e.g. in the warehouse).

(b) Documented information shall be adequately protected. A greater role than physical document damage due to improper use is data access authorisation and possible data theft. An excellent example of the non-fulfilment of this simple rule is the data theft by Bradley-Chelsey Manning and Edward Snowden. In both cases, highly sensitive data was easily accessible to an unauthorised user and caused extremely high damage from the point of view of the organisations concerned. To avoid similar experiences, sensitive documented information should also be password-protected and, where appropriate, subdivided into read and write permissions. A careful handling with IT-data is even more required, since operational values are bound ever more rarely in fixed assets, but increasingly in electronic documents and records. Their adequate protection is therefore existential. For this reason, the following should ideally be defined in writing: password rules, unauthorised software installations, access rights, procedures for destroying or decommissioning data media (hard disks, DVDs, etc.) and general instructions (or briefings) to sensitize employees to IT risks. In addition, security software (firewalls, virus protection, etc.) must be set up for data protection that is appropriate in quality and performance to the value of the data to be protected. Notes on data backup can be found at the end of this book chapter.

Besides all IT protection, a second focus on the protection of physical documents and records should not be forgotten. Sensitive data shall be protected from unauthorised access, e.g. by a lockable archive. In day-to-day routine, protection of documented information is also including, for example, the use of transparent sleeves or laminating documents in dirty environments or simply returning them to their intended place of storage if they have not been used for a longer period.

7.5.3.2 Control of Documented Information After Publication

After initial publication, the following requirements must be met in the course of documented information in accordance with Sect. 7.5.3.2:

(a) An internal announcement of new documents is necessary to ensure that the latest version of the document is always used in the workplace.

(b) For a structured procedure, distribution lists are suitable. These may need to be supplemented with briefings or training courses (e.g. in the case of new or significantly changed processes or technical adjustments).

(c) The *filing of* documented information must have a structure and order that makes it possible to retrieve data within a reasonable period. This applies in particular to day-to-day routine. Too often IT folder structures are not very logical and hardly comprehensible for outsiders or new employees. Moreover, folders for different projects in many companies are not structured synchronously and therefore cause unnecessary headaches for the users.

The *legibility* of documented information must also be guaranteed at all times. Documents must not be illegibly yellowed or dirty. In operational practice, the deterioration of condition is often a problem of records or of older design documents. The different storage media have different risks. Thermal paper, for example, tends to fade over time. In the case of electronic storage, file compatibility is a possible risk in the case of very long-term storage.

(d) *Changes to* documented information must be systematically controlled and remain traceable. Each company must ensure appropriate traceability of obsolete or no longer used documents and records. This standard requirement also includes the need for protection of documented information against unintentional changes.

(e) For the *archiving/storage of* documented information the following aspects are normally to be considered:[13]

- Controlled access to the archive should be ensured in order to minimise the risk of theft. It is also useful to monitor withdrawals of original documents and records and to trace responsibilities in the event of non-return.
- Records must be protected (e.g. from moisture, fire, theft) so that they remain legible during the prescribed period of storage.
- Retention periods should be defined. However, the standard does not provide any specifications in this respect. In this respect, it must be checked in each individual case whether customers, authorities or legislators (in particular via aviation legislation or national commercial law) specify deadlines. If this is not the case, a retention period of at least three years should be established. The EN 9130 standard may be a helpful guideline as it provides non-binding information on retention periods for records.
- An appropriate availability of archived documents and records must be ensured within a reasonable period. Appropriate means approximately within one to two days (within a certification audit).

(f) Particularly in case of new or updated documents, the collection of obsolete documentation must be controlled. In times of electronic documents this problem may have become smaller. However, there are still employees who store QM files as copies on the desktop screen or print documents before reading them and then store then storing the paper trays or drawers. They may still (unintentionally) use these versions even after a document update has long

---

[13]cf. in part also Sect. 7.5.3.2 b).

been published. Since this risk can never be completely excluded, the company should regularly draw attention to the associated problem[14] and thus create awareness.

Documented Information of External Origin

EN 9100:2018 requires that companies are not only responsible for their own documented information. It must also be ensured that appropriate control and storage of documents and records of external origin is set in place within the company's own control. An incoming inspection or a separate filing may be necessary.

Where documented information comes from external parties (usually customers or suppliers) and is not identifiable as such, it must be marked accordingly. It should also be noted that these must be treated with special care and confidentiality. In this respect, all requirements of Sect. 7.5.3 should apply equally to information documented both internally and externally.

Electronic Data Backup

In addition to Sect. 7.5.3.1 b), there are two explicit additional requirements for the handling of electronically stored data at the end of Subsect. 7.5. Organisations must create *data backups*. In small and medium-sized enterprises, regular systematic data backup sometimes shows potential for improvement. In those cases, appropriate attention has not been paid to data protection. It is often forgotten that operational assets are not primarily stored in (usually insured) buildings and facilities, but in IT. In this respect, it is important not only regarding the certification audit that the company has clear rules for securing the operational data. Backups must be stored in a different location than the original files. A fireproof safe in the same premises or a network backup at an external provider are just as suitable as outsourcing to another location (e.g. at the managing director's home or in a bank safe). The frequency of data backups depends on the size of the organisation and the data processed by the IT. Most organisations back up their data daily and weekly or monthly, and semi-annually or annually. It must be ensured that the back-up data is not overwritten during the next backup, otherwise permanent data loss may occur. If a loss of data in such a case is detected after three weeks of weekly data backup, this data has already been lost for two weeks. In this respect, longer data backups must be carried out regularly.

However, it is usually sufficient if the periodic data backup is limited to the transaction data. An incremental backup is also appropriate where only those files that have been changed since the last incremental backup are saved.[15]

---

[14]This can be ensured, for example, in an e-mail announcing the document revision. Such a message can end with standard phrase, e.g. "Please destroy printouts of the previous version". In addition, many companies point out in the footer of their documents that printed documents are not subject to document control and must therefore be destroyed after use.

[15]However, this of course requires that there is always a backup of the last full backup.

In the event of data loss or the non-availability (crash) of IT systems, companies should generally maintain an emergency plan or concept.

It is recommended to occasionally test the restoration of backed up data. This is often not possible.

As a side note, data backup also means that electronic documents and records (protocols, test reports, release certificates, approved design documents, etc.) can no longer be subsequently modified.

It is recommended – not only for the certification audit – to keep a documented overview of the IT landscape using simple means. This should provide information on the infrastructure, back-up procedures and, particularly for external IT support.

Within the framework of EN 9120, distributors are additionally required to keep information on the origin of the product, conformity and shipping.

References

CEN-CENELEC Management Centre: EN ISO 9004:2018-08 Quality management—Quality of an organisation—Guidance to achieve sustained success. Brussels (2018)

CEN-CENELEC Management Centre: EN 9100:2018—Quality management systems— Requirements for aviation, space and defence organisations. Brussels (2018)

CEN-CENELEC Management Centre: EN 9120:2018—Quality Management Systems— Requirements for aviation, space and defence distributors. Brussels (2018)

CEN-CENELEC Management Centre: EN 9130:2000 *Aerospace series—Quality systems— Record retention (draft)*. EN 9130:2000-09 Berlin Brussels (2000)

European Aviation Safety Agency—EASA: Acceptable Means of Compliance and Guidance Material to Commission Regulation (EC) No 2042/2003. Decision No. 2003/19/RM of the Executive Director of the Agency (2003)

European Aviation Safety Agency—EASA: Acceptable Means of Compliance and Guidance Material to Part 21. Decision of the Executive Director of the Agency NO. 2003/1/RM (2003)

International Aerospace Quality Group (IAQG 9100 Team): 9100 Revision 2016—Changes presentation clause-by-clause. May 2016 (2016a)

Operation

8

Chapter 8 of EN 9100 is focusing on the core elements of the operational value chain, namely design and procurement as well as production and service provision.

Chapter 8 of the standard begins with the planning of product realisation (Subsect. 8.1), progresses to the identification and evaluation of customer requirements (Subsect. 8.2) and continues (if applicable) with the design process (Subsect. 8.3). The provision of services is concluded with production or service provision that is executed according to clear specifications (Subsect. 8.5). The procurement is also defined as a core process (Subsect. 8.4). Chapter 8 ends with requirements for the release of products and services (Subsect. 8.6) and the control of nonconformities (Subsect. 8.7).

The core processes must be fixed in writing and controlled via KPIs and objectives (according to Subsect. 6.2). The type and scope of process measurement and target achievement are checked in each certification audit using the Process Effectiveness Assessment Reports (PEARs).

8.1   Operational Planning and Control

Long-term, high-quality service provision is only possible in an environment of clearly defined and systematically controlled processes. The operational value chain must therefore be executed under controlled conditions. Subsection 8.1 of the standard defines requirements which are necessary to implement a systematic framework for the identification and evaluation of customer requirements as well as for operational design, production and procurement. The structure and sequence of these value creation processes shall not be based "on a greenfield" without taking existing process into account. It must be planned and implemented in harmony with the other quality management processes (see Fig. 8.1 and 8.2). At the same time, the risks associated with the provision of the service must be adequately considered.

© Springer-Verlag GmbH Germany, part of Springer Nature 2020
M. Hinsch, *Guideline for EN 9100:2018*,
https://doi.org/10.1007/978-3-662-61367-2_8

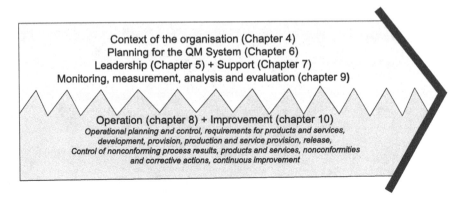

Fig. 8.1 Interlinking of product realisation and QM processes

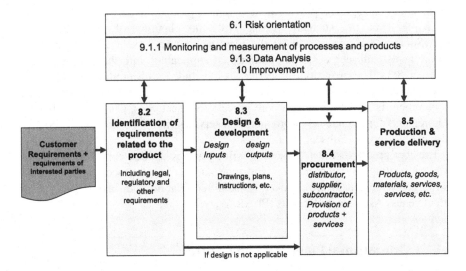

Fig. 8.2 Planning of product realisation

In order to establish EN-compliant value creation, planning elements and fundamental structures must be defined that are appropriate to the size of the organisation and the product or service portfolio. The standard provides explicit requirements for this which overlap with other chapters of EN 9100:[1]

---

[1]EN 9120 does not include individual requirements here and therefore has a slightly different sequence.

(a) *Product and service:* It must be ensured that the provision of the products and services takes place considering *all* requirements. The NOTE to a) indicates the type and scope of possible requirements in addition to those formulated by the customer, by authorities and legislators or by the company itself. The general standard requirement 8.1 a) is detailed in the Subsect. 8.2 and can therefore be neglected here.

(b) *Process specifications and test criteria*: It must be ensured that the operational processes are clearly defined, and the quality is sufficiently checked.

Processes: With the definition of the service provision processes, the framework of the value creation is defined. For this purpose, the company must determine how the process steps are to be carried out technically correctly. The individual activities are ordered and are placed in a process sequence that forms a meaningful value chain. The necessary process support must also be defined, e.g. through job cards and archiving systems, through IT support or by determining the scope of outsourcing. The level of detail of the process definition depends on the individual business case and the product or service portfolio.

Testing activities: For these, it must be defined when the process output corresponds to the defined parameters. The standard requirement 8.1 b) can be neglected here, as it will be outlined in more detail later. This is typical for design verification and validation (Sects. 8.3.4 c) and d) and 8.4.3.1), in production and service control (Sects. 8.5.1 c and g and 8.6) and in the course of monitoring and measurement (cf. also Sect. 9.1).

(c) *Resources*: Resources are essential inputs to the value creation and must be planned. Resources include operational production factors (personnel, premises, operating resources, IT) as well as products and services to be procured externally (e.g. materials, operating materials, components, temporary personnel, designs, etc.). As a result of the planning the necessary resources have to be determined and made available on time. It must be ensured that personnel capacity, technical equipment and premises are not only available in general, but on an order-specific basis in order to perform the service according to the requirements.

(d) *Process control*: The provision of services must not only be carried out in accordance with the defined processes, but also controlled and monitored. The type and scope of the monitoring are based on the specifications (see Sect. 9.1.1) so that the requirements can be skipped at this point.

(e) *Documented information*: Adequate documents and records must be available or created so that the added value can be appropriately instructed, carried out and demonstrated. This is a redundant requirement with Sects. 4.4.2 and 7.5, but also with further sections in Chap. 8, as documented information is explicitly required in the further course.

(f) *Handling of critical items*: A process must be defined for handling critical items. In the further course of the Chap. 8 of the standard, supplementary requirements for handling of critical items are found (cf. Sects. 8.3.5 e, 8.4.3 and 8.5.1 j), so that at this point a closer examination is not necessary.

(g) *Coordination with departments and teams*: In the course of (overall) planning and control of the provision of services, all respective persons responsible must be involved in the planning and control activities.

(h) *Elements for the use and maintenance of the product*: A discussion is not necessary at the beginning, because the requirements have been formulated in more detail in the standard's Sects. 8.3.5 (design outputs) and 8.5.5 (support after delivery).

(i) *Service procurement by external providers:* The scope and source of the procurement must be planned or controlled in close relation with the determination of resource requirements (8.1 c). In the certification audit, the original basis of the decision for an external service provision must be traceable, e.g. using planning tools or deriving them verbally.

(j) *Prevention of nonconforming products*: In particular Subsects. 8.6 and 8.7 are dedicated to these requirements in more detail. Therefore, a close look is not necessary at this point.

Outlocated Processes

Another part of Subsect. 8.1 deals with the partial or complete outsourcing of processes. The requirements for control and monitoring are based on the EN Subsect. 8.4 (Control of externally provided products & services). It is important to have the operational awareness that responsibility towards the customer cannot be delegated to third parties simply by outsourcing processes or parts of processes. The company must therefore ensure that customer requirements and other specifications are fully met when outsourcing. In a typical certification audit however, only the procedure for outsourced processes with a direct or indirect product reference are assessed.[2] At this point it is not necessary to take a closer look at outsourced processes, for details see Subsect. 8.4.

Project Management

Project management is defined as the organisation, planning, control and monitoring of complex one-off activities with a defined beginning and a defined end, accounting restrictions. Project management is explicitly required by the standard within in the context of product realisation, not for internal projects.

In addition to uniqueness, typical characteristics of project management are a clear time frame, financial and personnel objectives as well as limitations (e.g. resources), a project-specific organisation and a clear differentiation from other activities or projects.

From the standard's perspective, projects are expected to be handled in a structured, comprehensible manner and divided into work packages or stages. The core elements of project management are:

---

[2]If only activities not related to the product or service are outsourced, such as gardening work on the company premises or office cleaning, the standard does not require explicit control and monitoring activities.

- *Project preparation/approval*: Project preparation with project order, definition and objectives as well as project approval by the management form the basis of every project. Often however, only large companies acquire systematic project preparation in writing.
- *Project planning*: A sound project plan is essential for a stable and orderly executed project. The project plan describes the procedure of the project and determines what is to be done when and by whom. The depth of detail must meet the individual requirements of the project, so that both the project management and the project team can fulfil their tasks properly. The project is to be subdivided into delimited and manageable project phases with milestones. For this purpose, work/task packages are to be evaluated which are to be formulated clearly and realistically in coordination with the persons responsible for implementation. For the individual phases or work packages responsibilities and deadlines, input, expected results and any resource limitations must also be defined in addition to the content. Furthermore, scheduling as well as capacity and budget planning must be carried out in the planning phase. The volume of the project and the various personnel qualifications of the staff involved in the work packages must be determined for instance. A project-specific risk management system (cf. Sect. 8.1.1) must also be set up. Moreover, part of the project planning is the determination of the type and scope of the supplier integration. Since uncertainties are a part of everyday project management, the related planning is usually an iterative process.
- *Project realisation*: In the realisation phase, the focus is on processing the work packages in compliance with the plan and specifications. The project management controls the utilisation of resources. For the monitoring, which also represents a requirement, the target progress and target consumption are compared to the capacities actually used. The tracking of large work packages is ensured through strategic project monitoring (milestones/project review boards).
- Hereby the following tasks arise:
  - Evaluation and coordination of project bottlenecks and weaknesses (resources, costs, technical implementation),
  - Instructing necessary actions, and
  - Evaluation and approval of change requests to the project assignment.
  When deviations from the plan arise, actions must be taken, and the original plan updated in accordance with the new situation. In design projects, the review board members must approve the transition to the next project phase.
- *Project completion*. The project formally ends with the relieve of the project management by the customer, the review board or the management. Once the defined project goals have been achieved, the project ends with a project review by the team. The aim of such a de-briefing is to critically reflect on project execution, to identify improvements for future projects and to record these deficiencies in writing. Thereby a review also serves to fulfil the EN requirements for continuous improvement (Subsect. 10.3).

As part of a certification audit objective evidence for the planning activities, as described above, must be submitted. The type and extend of the documented information depend on the size and complexity of the project (e.g. project and milestone plans, feasibility evaluations, risk matrix, capacity forecasts and plans). For the realisation phase, it is primarily checked whether ongoing monitoring and control of target/actual differences is ensured. In addition, the auditor usually focuses on those procedures that deviated from the plan. A further focus is on checking whether project reviews have been carried out.

Planning of Work Relocations
If the location of the value chain changes, the associated relocation must be systematically planned and controlled. The change can take the following forms

1. in whole or in part,
2. temporarily or without a time limit,
3. through outsourcing,
4. by relocation within the own company or within the corporate group, or
5. by switching from one supplier to another.

For the standard it is irrelevant whether the work is outsourced to external suppliers, to affiliated companies, to other site of the company or simply to a different onsite production building or workshop. Both the transition phase and the restart must take place under controlled conditions. Particular emphasise is to be placed on the evaluation of possible impacts (risks) of the planned relocation. Both, for internal and external relocations, the contracted order or work package must be processed exactly according to the agreed specifications. In operational practice, planning deficits or insufficiently structured preparation as well as deficiencies in the communication of those involved often become apparent during relocations.

8.1.1   Operational Risk Management[3]

While Subsect. 6.1 deals with general company and organisational risks, this section of the standard deals with risk related with customer projects, orders and contracts. In contrast to Subsect. 6.1, methodical and systematic Risk Management (RM) is required here. This means that the procedure must be formally defined and described.[4] This can be defined in a separate risk management process. Alternatively, risk management can be integrated in an element of the production and project management process. It is the management that is responsible for ensuring that a process is a part of the order processing. Furthermore, through risk

---

[3]This section is not applicable to EN 9120 for distributors and stockholders.
[4]See EN 9100:2018 (2018), Annex A.4.

management the structured identification, analysis, evaluation and control of risks in projects and orders is ensured. The aim is to identify the risks in good time and to keep them under control or to eliminate them wherever possible by means of targeted measures.

At the project or order level, the risk process begins with the evaluation of customer requirements before the offer is submitted or the (internal) project assignment is approved. Once this step has been passed, risk management should always be firmly anchored in project reviews or other important project events, i.e. major purchases or outsourcing. At the project or order level, the project manager, production manager or account manager is normally responsible for ensuring a clear risk orientation.

The standard provides little information on the nature and extent of a project or order related risk management.

In order to provide a framework for risk management, risk activities must be considered in the individual process steps of order processing or in the individual project phases or milestones. For this purpose, risk aspects should be integrated into order acceptance forms, project applications, forms, checklists, production specifications or approval procedures.

The control of individual risks must be based on the potential extent of consequences and the probability of occurrence. In short: the greater the risks, the more extensive the measures to reduce the risk. This means for example, that the operational planning of risks of day-to-day operations (e.g. brief absence of employees due to illness) only have to be loosely monitored as they are associated with minor consequences. This is in contrast with projects and medium or large sized orders with specific requirements. In such projects, risk management must be evident throughout the entire order processing.

Within the scope of the certification audit, it must be clear that the risks are systematically recorded, from the acceptance of the inquiry right to the delivery (and possibly beyond). Moreover, the documented tracking of risks and if necessary, any required intervention in order to reduce the risks, must be demonstrated.

The following steps must be accounted for in a risk management process.

Risk Identification and Assessment
The first step in operational risk management is the identification of risks, e.g. the capacity or technical feasibility of customer requirements, potential design deficiencies, the loss of a key supplier or design partner, contractual uncertainties, investment or financing risks or uncertainties due to the use of new technologies. Since risks can therefore occur at most stages, the analysis must be carried out across all affected value-adding processes of the order. Only a broad basis makes it possible to identify all internal and external influencing factors and to gain a complete picture of the order-specific risk situation. The interactions between individual risks and any cumulations to major risks must also be considered. The recording is usually carried out via a risk register or an FMEA.

Following identification, the risks shall be assessed. Hereby the risks are to be classified in order to enable a targeted determination of risk handling. An exact measurement of the possible extend of damage or a precise determination of the probability is not a key feature. Risk quantification is usually difficult due to uncertainty in the assumptions. It is therefore important to form risk clusters. For this purpose, risk categories with criteria covering at least the consequences and the probability of occurrence must be defined, as this is the only method to enable standardised risk clustering. Afterwards, it is important to order the risks to clusters correctly according to their likelihood of occurrence to ensure an approximate risk ranking and prioritisation.

Risk Handling

Based on the risk identification and the subsequent evaluation, countermeasures must be developed, implemented and monitored. Although risk management activities depend on the individual case, there are four possible strategies that can either be combined or used individually:

- risk avoidance (avoidance of dangers while at the same time foregoing opportunities),
- risk mitigation (reduction of the risk to an acceptable level),
- risk transfer (complete or partial transfer of the risk to third parties, e.g. customer, supplier or insurance),
- risk acceptance (risk cannot be avoided, or the costs of risk mitigation are disproportionate to the benefits).

Countermeasures must be evaluated for their effectiveness. It must be checked, if the objectives of risk management have been achieved. If the remaining risk is unacceptable, new actions must be instructed.

During the certification audit it must be made clear that the risks are being addressed actively and with the intent of an effective PDCA cycle. For this purpose, records must be kept which demonstrate goals, deadlines, responsibilities and previous activities in risk management.

8.1.2 Configuration Management

Configuration management is the systematic control and complete documentation of product composition and properties over the entire product life cycle. Configuration Management (CM) thus serves the purpose of describing a product in its product design based on formalised procedures. Thus, a complete traceability of the technical product status should be possible at any time.

In addition to a clear product structure, the configuration includes all data and documents required for production, quality control and maintenance, such as specifications, circuit diagrams, drawings, parts lists, material requirements, test instructions, process specifications, program descriptions and digital mock-ups.

Based on this information, it must be possible to answer the following questions for each manufactured product at any point in time:

- What is the current product status? (In which physical construction status is the product currently?)
- How was the product developed? (Which design documentation is the product based on?)
- What influence does the product have on other components? (What are the effects of the product or configuration changes on other components and systems?)
- How was the product tested? (Which test environment, test parameters and test results formed the basis for the product release?)
- How was the product manufactured and in which status was the product delivered? (What was the construction status of the product at the time of delivery?)
- What changes have been made to the product since it was manufactured?

To meet these requirements the standard requires the implementation and maintenance of a structured documentation, testing and approval process. The design of this process is not explicitly specified. The EN 9100 however it does require to implement a configuration management that meets the requirements of the customer. The guideline for configuration management according to ISO 10007 provides non-binding information. This help document subdivides configuration management into the following four process steps according to Fig. 8.3

(a) CM planning,
(b) Configuration identification (incl. definition of the product structure),
(c) Change control, and
(d) CM-Documentation (Configuration Status Accounting).

Project + Number	
Customer specification	

Revision No.	Date	Name (Who?)	Description	Customer approval	Changes were reported internally
1.0	18.01.20	P. Miller	First Release	CEO, 25.9.2020	Engineering, Logistics
...	...	...	...	...	...

Requirement		Feasibility/ Risks/ Measures/ Proof				Release
Source	Requirement (Description)	Responsible for test	Comment/ Advise	Test carried out (Name / Date)	Fulfilled: Certificate/ Document	Fulfilled: Yes/No
...	...	...	...	...	...	...

Fig. 8.3 Core elements of configuration management according to ISO 10007

Effective configuration management begins with the definition of a framework.[5] In this context, the CM processes must be determined beforehand. The results of the CM planning can be a process description or a procedure or work instruction as well as the preparation of supplementary documents (forms, checklists, etc.). However, many companies describe their configuration management just roughly, because product status tracking is warranted by IT (in the background).

CM Planning

The definition of responsibilities, necessary control and monitoring activities and project or product specific documentation requirements are of high significance for planning. Furthermore, project management must be implemented for the design project. Attention shall be paid to a consistent enumeration system for documentation (partly based on recognised standards, e.g. ATA). This then serves as basic documentation structure and is applied throughout the entire product life cycle in all documents and records.

Due to decreasing vertical integration in the aerospace industry, it is becoming not only increasingly challenging for OEMs to keep track of design activities within their own company but also to external companies. Within the framework of CM planning, efficient interfaces to customers, partners and suppliers must be created. The type and scope of configuration management predominantly depend on the CM system (IT), but also on the complexity of the product. This allows OEMs to fully integrate their suppliers into their CM system or alternatively to standardise and actively manage the interface between the various systems. However, it is common for suppliers to receive little specifications regarding the documentation structure and designation from their customers.

Configuration Identification

The determination of main product groups (constituent assemblies) is the basis for the description of the product structure. Constituent assemblies describe by definition what is to be designed and controlled as a physically and functionally independent unit. For example, an aircraft consists of constituent assemblies such as the cockpit, engines, wings, fuselage, tail and landing gear. This definition divides the product into core elements in the top-level hierarchy. Therefore, the selection of the constituent assemblies should also be made at the earliest possible stage in the development process.

The constituent assemblies should be assigned so that they can later on be processed largely independently of one another. This is very important as the entire design complexity later on depends on this. The definition of the constituent assemblies significantly influences the transparency and thus the efforts for administration and control of the design project and therefore also the costs. This

---

[5]This is already formally part of CM planning. However, the process framework is usually anchored only once and is therefore explained here separately from the product-specific planning.

affect the design phase and subsequent phases. The selection of the constituent assemblies is influencing the manufacturing process and maintenance during the operation phase as well as during subsequent product changes. The definition of the constituent assemblies is therefore an important milestone for the entire design process!

The selection criteria to be considered when defining the main groups include the following:[6]

- product structure and product complexity,
- legal and regulatory requirements,
- customer requirements,
- criticality in terms of risk and safety (technically or economically),
- new or modified technology, design or development,
- interfaces to other configuration items,
- procurement conditions (in particular regarding location or supplier).

After the product structure is defined based on the constituent assemblies, the product is designed as so-called configurable items. At this second configuration level, the product is defined with all its properties and requirements. The technical development is then carried out on the basis of customer specifications, legal requirements, catalogues, manuals, specifications, drawings, etc.

As a result, a baseline (reference configuration) is created for the first time at an appropriate stage in the design-process. The baseline is an approved configuration status at a certain point in time, i.e. a configuration freeze. The baseline consists of the approved product configuration information that represents the definition of the product in detail. After the definition of the first baseline, any changes to the product configuration are to be monitored and documented. The currently valid configuration results from the baseline in combination with the design changes made afterwards. Usually, new baselines are set at least at the end of each important phase of the product design (milestones).

Change Control

A configuration is not a static structure but changes continuously with the design progress. Change management does not begin when the product has been fully designed and delivered to a customer for the first time. It already starts after the initial baseline has been defined. Even after delivery to the customer, change management plays an important role, e.g. in the context of modifications or major repairs. Configuration management must therefore include a continuous control system so that all changes are structured, coordinated, transparent and managed in a comprehensible manner throughout the entire product lifecycle.

---

[6]According to ISO 10007 (2004), Sect. 5.3.1.

When changes are made after the initial design, it is important that the original enumeration system is maintained and that the documentation specifications outlined in the CM planning are also applied to all phases of the product life cycle.

The CM change process is inseparable linked to the design change process in accordance with Sect. 8.3.6 of the standard.

CM Documentation (Configuration Status Accounting)
The last core element of configuration management according ISO 10007 is the standardised recording and archiving of records. The formats and structure of the records are based on the operational standards (CM plan, process instructions) and any customer requirements. The product related records primarily include all design data (drawings, circuit diagrams, parts lists, manufacturing and test specifications, etc.). Further records are used for clear tractability of the evolution of the developments and the decision-making basis. These are, for example, change requests, evaluations carried out, showing of compliance, minutes of meetings or releases and approvals.

The requirements of the EN 9100:2018 for configuration management are generally superficial. The standard merely requires the implementation of an appropriate configuration management system without mentioning details.

8.1.3 Product Safety[7]

EN 9100 certified companies must demonstrate a systematic approach to ensure high product safety. The aim of this requirement to put the products into circulation in accordance with principles that ensure their safety. In doing so, they should fulfil the promised or intended purpose of their product without endangering people or property.

Measures to increase product safety should focus on all phases of the product life cycle. In this respect, the requirement is focusing on all phases of the value chain, i.e. design, production, maintenance and purchasing, as well as on the period after delivery, i.e. product safety during operation and use.

From a standard's perspective, there is no need for an independent product safety process or even a safety management system. However, individual elements must be anchored in the value chain to systematically ensure product safety. This can be done, for example, by

- implementation of product and design-oriented risk analyses, e.g. FMEAs, safety analyses,
- consideration of double inspections in production,
- consideration of special testing and control measures for critical items,

---

[7]This section is not applicable to EN 9120 for distributors and stockholders.

- special incoming goods checks for selected products, services or suppliers,
- warnings for the user where human errors are to be expected or have already occurred,
- Poka Yoke, i.e. construction of technical precautions that immediately detect or prevent errors during use,
- improving the safety awareness of employees,
- identification and reduction of organisational risks and hazards in the area of human factors, regulation of responsibilities, which primarily achieves an indirect improvement in product safety.

The examples show that product safety measures are already being implemented in many companies. This is because even companies without a standard usually have a self-interest in high product safety.

In addition, implicit product safety requirements have already been formulated in other chapters of EN 9100, e.g. for special handling of critical items, risk management and foreign object control (FOD). It is important for a certification audit that an awareness and a structured handling of product risks can be demonstrated.

8.1.4   Prevention of Counterfeit Parts

Some prominent incidents in the aerospace industry, especially in the defence industry at the early twenty-first century led to a focused examination of the area of counterfeit parts. This has contributed significantly to the fact that this topic is now also anchored in the current revision of the standard. Counterfeit or suspected counterfeit parts are components, assemblies or materials that have been knowingly manufactured or maintained and released not in accordance with approved or accepted procedures or design data. Thus, the parts do not correspond to approved design data or the generally applicable norms or standards.

Counterfeit or suspected counterfeit parts may be identified for example by missing, implausible, invalid or falsified (release) certificates or accompanying documents. Previous occurrences, counterfeit markings, type plates, inscriptions or packaging placed may also be an indicator for counterfeit parts.

A key characteristic of counterfeit parts is that they were intentionally introduced to the market. This distinguishes them s from suspected unapproved parts (SUP) as counterfeit parts are of a fraudulent nature. These parts are classified as a risk by the aviation authorities. Therefore, their websites usually give information and assistance on identifying and reporting counterfeit and suspected unapproved parts. The FAA goes one step further and lists in its *Unapproved Parts Notifications* (UPN) those companies which have circulated unapproved parts.[8] During the last 10 years, only a small number of incidents per year were officially reported.

---

[8]cf. FAA (2020), http://www.faa.gov/aircraft/safety/programs/sups/upn.

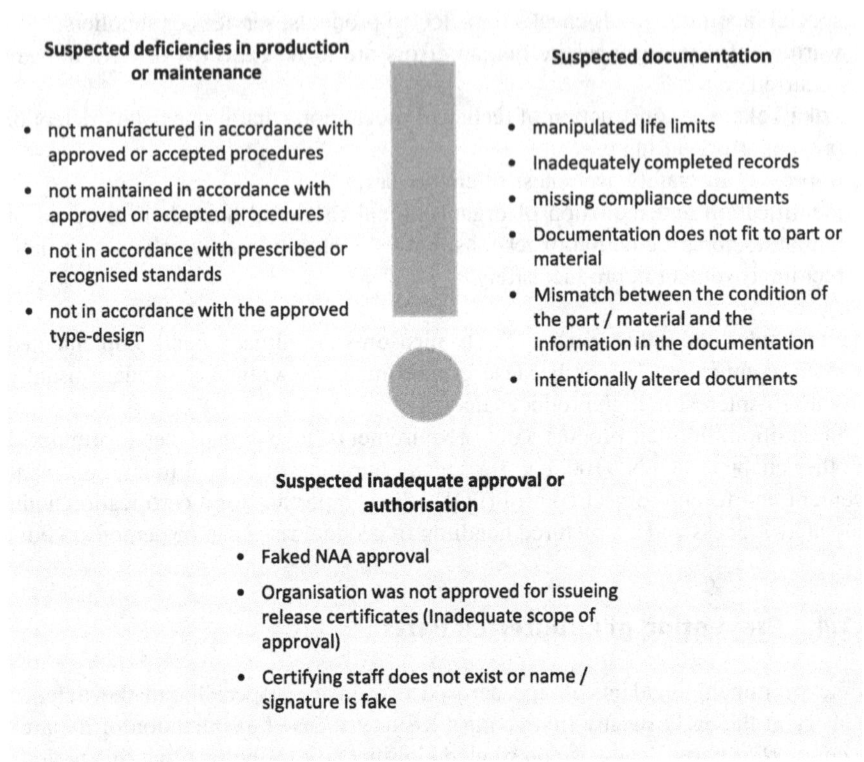

Suspected deficiencies in production or maintenance

- not manufactured in accordance with approved or accepted procedures
- not maintained in accordance with approved or accepted procedures
- not in accordance with prescribed or recognised standards
- not in accordance with the approved type-design

Suspected documentation

- manipulated life limits
- Inadequately completed records
- missing compliance documents
- Documentation does not fit to part or material
- Mismatch between the condition of the part / material and the information in the documentation
- intentionally altered documents

Suspected inadequate approval or authorisation

- Faked NAA approval
- Organisation was not approved for issueing release certificates (Inadequate scope of approval)
- Certifying staff does not exist or name / signature is fake

Fig. 8.4 Possible reasons for rejection of material or parts

Identifying these parts is not always easy in operational practice due to their high similarity to approved parts. In many cases, approved and counterfeit parts differ only in the manufacturing processes or the materials used. This is especially true for small electronic components.

Indications for unapproved and fraudulently placed part are, for example (cf. also Fig. 8.4):

- the prices requested or advertised are exceptionally low, they are far below those of the competitors,
- delivery times, that are significantly shorter than those of competitors, especially if the parts in demand cannot be delivered on the market,
- it is not possible for the supplier to supply drawings, specifications, manuals, detailed information on maintenance or on the certificates,
- during sales negotiation with the supplier, the impression is created that
  - remarkably large quantities of parts are available or
  - atypical payment methods are required (e.g. cash payment) or unusual (foreign) bank accounts are communicated for transfer.

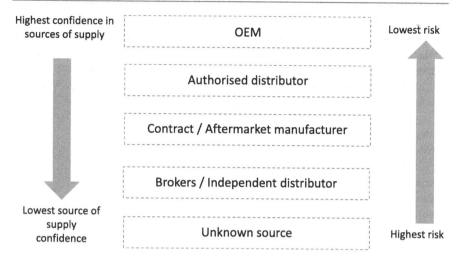

Fig. 8.5 Classification of external source risk

However, it should be noted that not all these indications immediately suggest a counterfeit part. These points are intended to help sharpen awareness. Conversely, counterfeit parts may exist entirely without these indications.

As a preventive measure against counterfeit parts in the own company, a careful supplier selection and supplier monitoring is required in a first step. In addition to its integrity and the experience with the supplier, it must be checked whether he has effective quality management systems that contain measures to identify counterfeit parts and prevent their use in operational practice (Fig. 8.5). Besides careful supplier selection and monitoring, the inspection of the parts is the most effective protection against counterfeit parts. In this respect, the utmost importance is given to the incoming goods inspection as well as to the employees working there, who must show an awareness towards counterfeit parts.

A prerequisite for the success of both measures is a solid training in the identification of counterfeit parts, especially for employees in purchasing and goods receiving.

Further measures to reduce the risk of counterfeit parts are:[9]

- Ensure complete traceability of all parts and systems right back to the OEM, other authorised manufacturers or the original recognised source of supply.
- Verifications and tests when dealing with parts of suspected or unknown origin, e.g. by means of parts identification by the OEM, performance tests or chemical analyses.
- Obtaining information from official warning lists, news from OEMs or other databases for counterfeit parts or suspected unapproved parts.

---

[9]The AS 5553 and the AS 6174 as well as the Supply Chain Management Handbook (SCMH) of the IAQG offer helpful information for the prevention of counterfeit parts.

According to Subsect. 8.7, counterfeit or suspect counterfeit parts must be treated separately and closely monitored in order to prevent accidental backflow into the regular material cycle. When such parts are identified the customer must also be informed about these parts. Companies with an EASA approval according to Part 21G or 145 must also inform the responsible national aviation authority. A procedure for handling identified counterfeit parts includes the following steps:

- identification of all locations of the material,
- blocking of all associated warehouse stocks,
- clear identification of suspicious parts,
- separate storage to keep the affected parts out of circulation,
- gathering objective evidence,
- notifying the supplier and, if applicable, the customer,
- root-cause/origin analysis,
- scrapping.

Note on EN 9120: The standard for distributors has an additional section, Sect. 8.1.5, where a process to identify and prevent suspected unapproved parts is outlined. These parts must be considered to be counterfeit until their origin is clarified. However, there is a high level of redundancy in regard to Sect. 8.1.4 which describes a process for handling suspect counterfeit parts.

8.2   Requirements for Products and Services

8.2.1   Customer Communication

In this subchapter, the standard requires companies to establish adequate communication structures with the customer. Customer communication can essentially arise from three sources:

- non-specific communication with customers or potential customers, e.g. regarding general product information, marketing, general communication standards,
- communication with regards to a specific request or related to an order,
- communication providing feedback, for example suggestions for improvement, recommendations or complaints as well as positive customer response.

Communication can take place through personal conversations or letters, via telephone, email, document exchange, IT platforms, portals or customer events. Since communication comprises all forms of information exchange, marketing channels such as information and advertising material (brochures, flyers, technical data sheets, etc.) and the company's own website are also regarded as channels of customer communication.

The standard gives few indications as to when customer communication or related organisational structures are appropriate in extent and nature. This is mainly due to the fact that customer communication depends to a large extent on the type of service provided and the significance of the customer. As a rule of thumb, the communication structures can be described as appropriate if it can be assumed that the customer feels adequately informed, considering its significance for the volume of business. For this purpose, customer communication must ensure that:

(a) there is an appropriate exchange of product or service characteristics.
(b) there is sufficient coordination with the customer when the order is initiated and concluded. This applies in particular to adjustments that are made in the period between the first contact and the conclusion of the contract. In addition, changes in the scope of services must be communicated that are required or necessary during the contract period.
(c) customer feedback is systematically recorded and processed by the organisation (and not only by individual employees). This applies especially to customer complaints.
(d) requirements for handling of customer property are agreed at an early stage where applicable.
(e) where applicable, emergency procedures are defined (in particular in case of failure of resources such as equipment or IT systems).

Good communication is actively embraced so that it is not just a customer initiative but is actively anticipated by the company.

In the certification audit, the focus is often on communication in the context of customer complaints and on coordination during order initiation or during service provision when changes occur.

8.2.2 Determining the Requirements for Products and Services

Knowing what the customer wants is a prerequisite for initiating a business relationship and consequently for delivering the agreed product or service to the customer. Customer needs must therefore be identified and realised if the company is to meet the core objective of EN 9100, namely the customer satisfaction.

The requirements of this section of the standard focuses on the ability of the company to determine and meet customer requirements. The focus here is on the process capability to deliver the products with the announced requirements. This includes the identification and ability to meet regulatory, legal and operational requirements. The standard is intended to prevent immature services being offered with unfulfillable requirements in advance.

Since this section of the standard deals with the determination of product and service requirements, the fulfilment of these requirements is necessary before entering a delivery obligation.

Project or order specific risks (delivery times, new technologies, etc.) must also be identified and managed, e.g. by analysis and evaluation of the following information:[10]

- customer requirements,
- own strategy and business plans,
- product analysis according to the Kano model,
- market observations and market surveys,
- design activities,
- customer feedback,
- new or changed regulatory or legal requirements,
- new technologies or changed process technologies, and
- feedback from suppliers.

8.2.3   Review of the Requirements of Products and Services

The determination and evaluation of the requirements for the product or service is carried out in day-to-day operations before the contract is concluded. Thereby, this process is the responsibility of the sales department or sales related departments (e.g. customer service).

In the simplest case, the company receives a part number from a potential customer in order to submit a quotation. For more complex products and services, the company receives the order requirements in the form of a specification. The expected requirements regarding functionality, design, performance, materials as well as qualification and acceptance are then formulated in this document by means of written descriptions, lists, drawings, and photos. In most cases, the enquiry or order is accompanied by additional requirements regarding packaging, transport and delivery or maintainability. Finally, customer needs with regards to order processing such as delivery dates and responsibilities are specified. Based on this information provided by the customer, a conclusive and unambiguous a description as possible of the service to be provided should be created.

Complex customer enquiries must first be broken down into individual requirements. A systematic evaluation can only by ensured if requirements have been individually determined beforehand.

Sales or customer support often only hold a coordination and interface function between the customer on one hand and the specialist departments on the other. This particularly applies to large or complex inquiries. Here, not the sales department does the main the main work in identifying and evaluating customer requirements and organisational needs but engineering, work planning, production management, materials management and controlling as well as suppliers do. In

---

[10]See ISO 9001 Auditing Practice Group—Guidance on: Customer Communication, (2016), P. 2.

Project + Number	
Customer specification	

Revision No.	Date	Name (Who?)	Description	Customer approval	Changes were reported internally
1.0	18.01.08	P. Miller	First Release	GF, 25.9.2017	Engineering, Logistics
...	...	...	...	...	...

Requirement		Feasibility/ Risks/ Measures/ Proof			Release
Source	Requirement (Description)	Responsible for test	Comment/ Advise	Test carried out (Name / Date)	Fulfilled: Certificate/ Document	Fulfilled: Yes/No
...	...	...	...	...	...	...

Fig. 8.6 Example of a compliance matrix

addition to product design and product or service characteristics, important evaluation criteria are for example competence, manufacturing processes, materials used, capacity, delivery times and price expectations.

Strictly speaking, a review of the requirements relating to the product or service consists of two steps: the identification and the evaluation of the requirements.

Identification of Requirements Related to the Product or Service

The compliance matrix is an instrument for recording customer requirements as well as additional legal, official or other requirements (see Fig. 8.6). These include a short description, significance of the requirement, internal responsibility, special characteristics and risks as well as open items or criteria for demonstrating compliance. Ideally, there should also be a column indicating who was informed in case of any changes to the requirement. A compliance matrix is a living document and is completed successively. In the first step, the identified requirements are typically listed, while in the second step, a review or an evaluation is made along with the documentation of detailed information.

When entering customer requirements, it is important to ensure completeness.[11] The customer's specification is not always complete. For example, individual requirements may have been forgotten by the customer due to lack of experience, ignorance or carelessness, or they may be implicitly assumed (e.g. CE certification).

Further requirements may arise from interested parties which have not been formulated by the customer. These do not necessarily have to be official requirements; they may also be requirements of end customers (e.g. passengers, crew,

---

[11]For information on the various types of requirements, see Sect. 8.2.3 a)–c) and the NOTE.

maintenance organisations). Finally, requirements must be considered that arise from the company's own quality standards.

In case of doubt or ambiguity in the identification of individual requirements, the customer should be consulted.

For mass-produced goods such as standard parts, the determination of product and performance requirements is less detailed than for non-standard products. In this case, customers order articles through defined part numbers from websites, brochures or portals, so that the company only has to identify the ordered items, the delivery date and any special terms of delivery.

Evaluation of the Requirement with Regard to the Product or Service
Once the requirements for the product or service have been determined in a first step, they must be evaluated in a second step. This should answer the question if the company can fully meet the necessary requirements and if any uncertainties and risks exist in the context of the potential order. This needs to be clarified,

- before submitting a binding offer, or
- before entering a delivery obligation (contracts or orders) or
- before acceptance of contract or order changes.

For the assessment of complex services, the existing compliance matrix is completed, line by line, requirement by requirement. The focus here is on the documentation and the evaluation of operational resource availability, delivery dates, uncertainties in the design or in the manufacturing processes.

Risks and requirements for demonstrating compliance as well as other open items must also be documented in the matrix. This is especially important for major orders or new customers. New technologies can represent a risk, with regards to process technologies, process structures, software or products used for the first time in operation and with regards to new technologies available on the market (e.g. 3D prints or new composite mixtures). For risky customer inquiries, the related risks must always be determined individually, evaluated and mitigated with specific solutions. Here there is a close tie to operational risk management (Sect. 8.1.1).

The evaluation of product and service requirements shall also include the availability of capacity and, where appropriate, at least a rough project or order planning. Both are important because this is the only mechanism to determine whether the sales order can be fulfilled at the expected delivery time.

If the technical feasibility result is positive, a consolidated commercial evaluation must follow. For this purpose, external services as well as direct material costs (incl. special procurements, training) are to be identified besides labour hours and costs. Expenses that can only be indirectly assigned to the order (overhead) are also to be considered. A sound process for order calculation is important from a standards perspective as this is the only instrument for ensuring a long-term operational market presence.

In a certification audit, it must always be expected that the documentation for the technical and capacitive evaluation of a customer inquiry will be checked. The identification and evaluation of order specific risks is also an aspect of every sales audit. In this respect, it is important that appropriate records are kept of the identification and evaluation of customer requirements as well as the associated measures.

Submission of a Proposal

For the finished offer, it should be ensured that handwritten comments have been incorporated and that there are no loose notes. The offer should be presented in an orderly, understandable structure. In addition, once all the decision makers involved have checked their offer and approved it by e-mail or signature, the offer can be submitted. A signature regulation for sales activities should be in place in order to reduce risk.

If the offer has been accepted by the customer, the order confirmation or a subsequent contract must be checked for conformity with the offer. The delivery obligation may only be entered if the information in these documents are consistent and any deviations have been clarified.

If the customer has not or not sufficiently specified his requirements in writing, the company must do so in the proposal or in the order confirmation. This is intended to provide both parties with a clear picture of the scope of the contract and to avoid potential conflicts at a later stage.

For mass produced goods or standardised services, another focus of evaluation is to be applied. After all, a detailed individual check of the feasibility of a customer request is neither helpful for the customer nor for the company. In this case it is more appropriate to pay attention to an adequate and up-to-date description of the products and services through sales. The evaluation should then focus on a comparison of the order data with this sales information and the requested delivery date.

8.2.4   Changes to Requirements of Products and Services

Large, complex proposals and orders are developed in an iterative process. Until the contracting parties finally reach a delivery commitment, they have often iterated through several coordination loops with numerous changes. To ensure that all demanded adjustments are incorporated in the (final) specification or contract, it is important that changes are immediately incorporated into the documentation and the requirements of previous drafts are revised. The documents are usually stored in an electronic form in a project or quotation folder, in which the latest version of the complete documentation is stored. Changes have to be communicated to those involved in the company in order to create awareness of the latest changes.

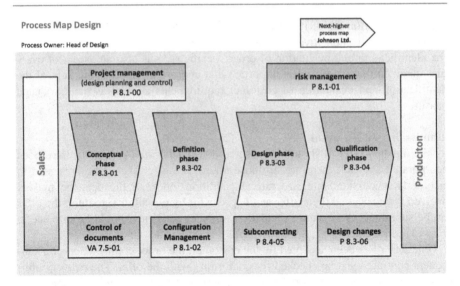

Fig. 8.7   Example of a design process map design

8.3   Design and Development of Products and Services[12]

8.3.1   General

At the beginning of every product life cycle there is a design that serves to transform an idea into a marketable product. After the market launch, design activities again play a role when modifications, extensions or extensive repairs are made to the product.

Companies that include design services in their range of services must carry these out under controlled conditions and therefore implement and apply a design process.

Design projects must therefore have appropriate organisational structures to enable full and effective cooperation between the parties involved. A written design process must be defined for this purpose. If the design process consists of several individual processes, it is advisable to provide them with a process map *design* (see e.g. Fig. 8.7). This not only helps the auditor, but also creates a comprehensible working basis for employees and an awareness for the embedding of their own tasks within the overall process.

Section 8.3.1 contains no specific requirements. If the organisation is executing designs activities, the requirements of Sect. 8.3.1 are fulfilled as soon as all other design requirements (Sects. 8.3.2–8.3.6) have been implemented.

---

[12]This subchapter is not normally applicable to EN 9120 for distributors.

8.3.2   Design and Development Planning

The capacity and financial resources of most design projects are economic and scheduling objectives can only be achieved through a systematic approach. For this reason, companies must plan and monitor their design activities in a structured manner. For this purpose, project management is used in operational practice. This includes design planning, implementation and monitoring. The starting point for this is usually a project assignment or a customer order on which basis a project plan is drawn up in which the design project is structured and laid out comprehensibly. Such a plan must have a level of detail that enables the project status to be determined at any time (in particular target/actual hourly consumption, achievement of project milestones, status of project risks).

In addition, the configuration management (CM) plays an important role in design planning because it directly intervenes in the design process by defining a hierarchical product structure. The CM is therefore a sub-process that runs continuously alongside the entire design and project management process. In operational practice, design planning and configuration management (planning) are therefore closely related.

(a) The individual design stages must be formulated comprehensibly regarding scope, task and objective. Furthermore, the expected results must be defined. A project plan provides an overview of this, showing what needs to be done when and by whom.

(b) In order to have a clear structure, design projects must be divided into phases. These are derived from generally recognised project phases as well as from project-specific work packages. At the end of each phase there are management-oriented design reviews as well as operationally oriented verifications and validations of the outputs (cf. Sect. 8.3.4—Design control).

(c) It shall be determined when verifications and validations are to be carried out. Often this already results from the definition of the project phases (see previous bullet point). Requirements for the implementation of verifications and validations are contained in Sect. 8.3.4.1.

(d) Design phases and work packages must be assigned to departments, teams, suppliers or individuals. This applies not only to operational design tasks, but also to management activities, such as responsibility for approvals, e.g. by the design team, the project board or the customer.

(e) Resources must be planned realistically. In addition to man-hours in engineering, this includes resources and material costs in production of prototypes, qualification and project management. If external engineers are involved to a design project, these expenses must also be considered in the design planning. Certification audits show that the planning quality of design projects in daily practice is sometimes insufficient. Too often, very tight calculations are made, without necessary buffers in terms of deadlines and capacity.

(f) All teams, departments and parties involved in design project must be appropriately involved in planning and controlling. In this way, all disciplines can contribute their requirements and experience. The comprehensive involvement of all stakeholders is not only related to internal interfaces, but also at suppliers, customers and other interested parties (e.g. aviation authorities).

(g) The standard explicitly points out that customers and other users may also be involved in the design process. In operational practice, it is not uncommon for the customer to participate not only in the (strategic) design reviews. Often, he is part of operational coordination meetings in order to monitor the course of the project and to instruct actions at an early stage in case of problems.

(h) When planning, it must be explicitly considered that the later manufacturability and maintainability of the product is adequately ensured within the scope of the design activities. If no or too little consideration is given to this, this can lead, for example, to
  - unnecessary production complexity o
  - high product susceptibility due to incorrect material selection,
  - high maintenance efforts, e.g. due to poor accessibility. The result is a lack of customer satisfaction or a lack of customer interest right from the start. Design deficiencies can therefore always have an impact on the long-term economic success of a product.

(i) The planning and control activities must be of a scope that customers and other parties (e.g. the certification auditor) consider appropriate. This requirement is fulfilled from the customer's perspective if he does not formulate any improvements to the project planning and control. From the standards point of view, this requirement is met if the previous enumerations in this section have been checked without any findings.

(j) Documents and records shall be kept for design planning and control so that it can be comprehended that the design project (management) was carried out to an appropriate extent.

For small or simple design activities (e.g. change notes/requests), the planning and systematisation of the work can be ensured by using forms or an IT workflow instead of extensive process and project structures.

Design Planning in the Certification Audit
In a certification audit, a current or recently completed design project is usually inspected. The focus is initially on the project plan that must be explained by the audited employee. In this context, the planned design resources and those already used must be shown in addition to internal working hours, which also include external services and any material costs.[13]

---

[13]A compliance matrix can be used to determine expenses for example. By using a bottom-up approach man-hours, external services and material for the individual requirements can be estimated.

Overall, it must be evident that the design project is conducted under controlled conditions. It is therefore to be expected that the auditor will ask the following questions regarding the design planning:[14]

- How is the design process and planning generically structured and described (detached from individual design projects)?
- Which resources and competencies are needed for the design project, when and to what extent?
- How are responsibilities defined? How is the contact with the responsible 21 J design organisation or the EASA planned?
- Which services are to be subcontracted? What does the corresponding detailed planning comprise (work packages, milestones)?
- Which internal and external interfaces have been identified between the different teams? How is appropriate coordination between these parties ensured?
- How are subcontractors involved in the overall planning and how are they later controlled?
- Are checkpoints for verification and validation adequately defined?
- Are the schedule and milestones realistically planned?
- Are appropriate activities/measures planned to monitor the effectiveness of the planning?

In the certification audit the connection between design planning and the

- risk management (Sect. 8.1.1),
- the configuration management (Sect. 8.1.2),
- supplier management (Sect. 8.4.2),

will typically be closely examined.

The company must be able to demonstrate documented risk activities for each design project.

8.3.3 Design and Development Inputs

Design activities are initiated from documented guidelines and verbal information which provide information on what is or should be designed.

Inputs form in their entirety a description of the planned design. In order to get a comprehensive and accurate picture of the requirements, the inputs must first be defined and gathered. This takes place predominantly in the tendering process

---

[14]Similar: ISO 9001 Auditing Practice Group—Guidance on: Design and Design Process, (2016), P. 3.

(cf. Sect. 8.2.3). The aim is to obtain a complete, conclusive, consistent, unambiguous and functional description of the product or service requirements.[15]

Further inputs are qualification requirements (e.g. reliability, reaction speed, look & feel, tolerances, weight, cleanliness) as well as specifications for costs, data protection, maintenance, material or transport and storage. When gathering the inputs, it is useful to draw on lessons learnt from comparable design projects in the past. Generic checklists or previous specifications may also help to ensure that important input is not forgotten.

Design inputs result from written or expected requirements of the external or internal customer. Furthermore, operational specifications, industry standards and voluntary commitments can be additional inputs. In accordance with the listing in EN Sect. 8.3.3, legal and official regulations must also be considered when collecting input data for a design.

Typical documents for the design are therefore:

(a) customer specifications with functional and performance data, results from market analyses,
(b) (documented) experience from previous design projects and data of products and services already on the market.
(c) EASA Certification Specification (CS) as well as Guidance Material and AMC,
(d) recognised design and qualification requirements, industrial standards or other generally recognised process standards,
(e) risk analyses with regard to potential errors or types of errors and their effects on the products and services to be designed. These must then be prevented from the start by skilful design. An example is the consideration of a construction about dimensioning, material fatigue, breaking strength or the analysis of possible consequences with regard to fire hazards, system failure, electrical radiation.
(f) Investigations, analyses and other documents that allow conclusions to be drawn about risks regarding the long-term availability or interchangeability of material or documents. The standard does not call for a dedicated obsolescence management. However, it is necessary that availability aspects are explicitly considered in the design.

When changes to existing products or services are incurred, the design inputs include change requests from the customer, observed or reported quality defects and improvement potentials as well as self-identified supplementary requirements (in-house or through market studies, etc.).

---

[15]For physical products, the fundamental functional and technical requirements subsumed under the so-called "4 F" (form, fit, function, fatigue as well as qualification) can be used as a mental checklist.

Records shall be kept of the design inputs.

It is important that the design inputs are structured and fully and precisely defined to form a picture that is as clear and uniform as possible. For this purpose, different requirement levels can typically be distinguished in operational practice:

- *Must* or mandatory criteria: indispensable characteristics of the product, where it is compulsory to ensure compliance,
- *Should* criteria: compliance is not directly necessary, however, a realisation of the requirement is desired,
- *Shall* criteria: the fulfilment is not necessary, but it is targeted, if it does not exceed the planned use of resources,
- *Exclusion* criteria: these requirements explicitly outline that certain criteria will not be met (exclusion principle).

8.3.4   Design and Development Controls

Design control is about monitoring the status of (technical) product development and project progress. This is to be ensured through a formalised procedure in which systematic checks are continually made against design planning and inputs. If problems are identified, measures must be instructed, implemented and monitored via the development control.

Design control is not only an operational task, but also management job.

(a) *Definition of design outputs*

The design outputs are to be defined with regards to the (technical) content as well as to the capacitive and time related project requirements. The type and scope of the design output is roughly based on the design inputs, e.g. it is based on the customer specifications. From the perspective of the EN 9100, it must be ensured that the information is complete, comprehensible and understandable, as well as correct and consistent.

As soon as a design project has been started, it must be controlled not only technically in terms of design and construction, but also in terms of capacity and schedule. In this respect, the objectives of the design projects also include requirements in the area of project management (e.g. adherence to deadlines, costs, man-hour consumption).

(b) *Design reviews*

Within the framework of design reviews, the aim is to systematically evaluate the status and course of design project against the design planning and design inputs on a higher level. The design review as defined in the standard, Sect. 8.3.4 b) is a management task. Typical design evaluations are, for example, the following

- Preliminary Requirement Review, in which a decision is made on the feasibility of the project and the submission of an offer.
- System Specification Review for which the specification (requirements) is defined and released.
- Preliminary Design Review, for which a rough draft of the design has been worked out and approval for detailed design is being given.
- Critical Design Review, during which the final design review is carried out and a decision on a design freeze is made.
- Verification Review for which the showing of compliance has been completed and the release for (series) production is being approved.

During the reviews it is hardly possible, especially with larger projects, to check each requirement individually for its fulfilment. However, this is not usually the purpose of such review and it is generally not intended to achieve this. After all, such reviews are usually scheduled for two to four hours and often take place with the participation of 5–10 managers. It is primarily a matter of determining the general project status and the progress of the design results. In addition, important decisions, possible problems or risk potentials are to be addressed and any necessary actions instructed. Design reviews should include managers of the design departments and the heads from some of the following areas: Purchasing, sales, manufacturing/production planning, quality management and customer service. It may also be necessary to involve the customer. Due to the wide range of participants, a diverse range of experiences and opinions can be used. This is the best way to identify and address risks and problems as well as potential for improvement at an early stage. In a certification audit, the following questions of the auditor can be expected:[16]

- Have reviews been carried out at planned times/milestones in the design project?
- Are all reviews carried out in a systematic, comprehensible manner with all important coordination aspects?
- Are all teams, groups or departments that were involved in the respective design phase represented in the reviews?
- Have all original and any new design inputs and outputs been addressed or considered?
- Have design changes been approved by those responsible (including the customer, if applicable)?
- Have the planned goals of the design phase been achieved and has the progression to the next phase been authorised during the review?
- Are there enough records of the review?

---

[16]Similar ISO 9001 Auditing Practice Group—Guidance on: Design and development process, (2016), P. 5.

While strategic design control is carried out via design reviews with management, daily and weekly project monitoring is in the responsibility of the project manager and, if necessary, other sub-project managers. Operational management focuses on the ongoing monitoring of the degree of completion of individual work packages. The planned progress of work is compared with the capacities used, usually based on the hours booked by the employees on the work package. Such a comparison is usually carried out on the basis of daily or weekly actual and target values per work package level. This enables a quick identification of deviations from the plan and the initiation of countermeasures. Design activities must be instructed in a form that enables the executor to recognise which requirements and objectives must be fulfilled.

(c) *Design verification (see also* Sect. 8.3.4.1)

When at a design phase or at the end, when all design activities have been completed, the design solution must be verified. With these reviews it is checked, if the design output corresponds to the specifications and all other inputs. Verification is therefore a check against the originally planned design requirements (have the engineers fulfilled the design objectives defined in writing beforehand?). The operational *responsibility for* design verification cannot be delegated to suppliers and especially not to the customer!

The verification focuses on a technical verification regarding the fulfilment of the design requirements, e.g.

- the design and execution (form, fit, function),
- reliability,
- properties such as strength, flammability or load/resilience, safety, comfort,
- the operational behaviour, i.e. the performance and operational characteristics and limitations,
- the Look & Feel or
- the cost/price.

In this context, the design assumptions (e.g. load, heat, electromagnetic compatibility) should be checked. Methodologically, the following methods can be used for verification:

- document checks,
- calculations and analyses,
- simulations and
- inspections and/or
- tests.

In addition to technical verification, formal aspects of the documentation should also be checked, e.g. completeness, correctness and plausibility as well as compliance with operational and industry standards.

Verifications must be planned. In many cases, the (customer) specification gives hints for the definition of verification methods. If this is not the case, the verification specifications, such as test procedures and acceptance criteria, must be created or completed at the latest in the course of the actual design activities (see also Sect. 8.3.4.1).

The verification is a second check during which another employee inspects the work. The latter must not be directly involved in the direct design activities to be examined. However, membership of the same department or team is permitted. It must be ensured that the employee responsible for verification is qualified for the task.

Verification can be performed separately or in combination with validation.

(d) *Design validation (see also* Sect. 8.3.4.1*)*

Following or parallel to the verification, the validation takes place. While the verification checks against the specification, the validation checks against the original purpose (e.g. customer) as well as against official or legal requirements.

Based on the design outputs (e.g. calculations, analyses or with a sample component of the so-called Qualification Unit), the design validation is conducted by the company, however, mostly on-site in a system-integrated status after delivery at the customer's premises.

The customer is then provided with the design outputs and, if necessary, a qualification unit together with the associated test equipment. In this case, the company only provides technical support. The methods of validation may correspond to those of verification. In addition, these can be pilot projects, field studies, tests on prototypes and/or tests on system-integrated sub-assemblies. The validation methodology is often already derived from the customer specification, otherwise directly or indirectly from regulatory Certification Specification (CS or FAR).

All validation activities and results shall be documented. If the validation is carried out by the customer, it is not unusual that the customer does not provide the company with any documents and records of his testing activities. In the context of certification audits, however, missing validation records are normally not a problem.

If the design outputs have passed the validation, the design process is usually completed, and a design freeze is authorised. This is the basis for a production release. The authorisation is to be made in writing for reasons of traceability. If agreed or appropriate, the customer shall be involved in the decision.

Neither the verification nor the validation necessarily has to be at the end of the design activities. This usually makes sense, however, there can also be partial exams on the incompletely developed product. This is the case, for example, for test points that are no longer accessible for installation or for quality tests (fire tests) with materials that are to be installed. Intermediate reviews should also be carried out if the continuation of design activities with incorrect data results in unreasonably high costs.

(e) *Necessary actions in case of problems and deviations*

If problems or risk have been identified during design reviews, effective counter-measures must be taken. In any case, there is a need for action if the results do not meet the requirements and if deviations in the project planning (hours, costs, schedule) have arisen or are to be expected. For example, a re-design is necessary or adjustments in the project or resource planning become necessary. The processing of the identified problems and risks must be tracked in a comprehensible manner. Their status must be checked or discussed in the next reviews.

(f) *Documentation of design control*

Documented information shall be prepared on the activities of design control and design verification. These shall be procedure or records. In order to facilitate verification and avoid forgetting test criteria, it may be useful to use checklists. In addition, the use of forms contributes to standardisation when recording test results. This increases transparency and error reduction.

(g) *Authorisation of design phases*

If, in the course of the design reviews, all design objectives of the respective phase have been achieved, the progression to the next design phase shall be authorised. In operational practice, a formal and comprehensible release is sometimes forgotten and leads to a finding in the certification audit. The results of design reviews shall be documented. In many companies this is done by means of minutes of meetings.

8.3.4.1 Design Verification and Validation by Means of Tests

This section is a specification of the previous Sect. 8.3.4 c) and d) for design verification and design validation. If tests are used, these must be systematically prepared, carried out, evaluated and documented.

As part of test preparation, test plans, specifications and/or test procedures must first be created or provided by the customer. The preparatory test documentation must contain the answers to the following questions:

- What should be tested and why should the test be performed?
- How should testing be performed (test setup) and which test parameters are defined?
- What is required for the test (test environment)?
- At which point in the process should the test be carried out?
- How are test results to be evaluated (e.g. acceptance range)?

Part of the test preparation is, besides documentation, the provision of adequate and sufficiently calibrated test equipment and measuring instruments (in accordance with Sect. 7.1.5) as well as the ensuring of acceptable environmental conditions. Finally, care must be taken that the test is performed against the correct configuration status of the product to be tested.

All results of the design verification and validation must be traceable. In order to demonstrate that the product meets the requirements of the specification, the following planning and execution must be documented in addition to the design outputs. Thereby, the entire verification and validation process, as well as the derivation of the results, can be explained. The following documents and records must be kept for this purpose: Test specifications, test plans, test procedures, test reports, completed checklists, acceptance protocols and releases.

The results of the test are recorded in test reports. The acceptance criteria (pass/fail) previously defined in the test plan or test specifications are confirmed (or rejected) in the test reports after test execution.

In the test documentation, care must be taken to ensure that measured values are recorded if tolerances are specified (e.g. 12.2–16.4 V). An "OK" or "FAILED" is not enough. In the event of a failed test, a documented root cause analysis must be performed (cf. Subsect. 10.2).

8.3.5   Design and Development Outputs

At the end of each design project there must be results in the form of technical documents and, if necessary, models or prototypes. This output must not only be clear and comprehensible but must also be provided in a form that can withstand a comparison with the original requirements (design inputs). In the end, it must be ensured that the design results meet the design specifications.

The design results must show a degree of detail with which it is possible to produce or execute the design in constant quality without further technical queries. The results can be specifications for production and for maintenance, but also for use or operation. When drawing up design documents, operational or industry-specific standards shall be used (cf. Sect. 8.3.3), e.g.:

- Requirements for the format and structure of the design documents,
- Reference to standard procedures instead of own specifications,
- Use of forms, text modules and simplified English.

Of course, documented information on the design outputs must be created. However, for most design organisations, this is a matter of course; design activities worthy of the name, without the creation of documents, are likely to be an exception even if a company does not have an EN certification.

An important part of the design output are production and maintenance specifications. This data includes all information that describes the work to be performed, procurement and testing of the product or service. These documents are, for example:

- Specifications, drawings, calculations, assessments, photos, layouts, drafts, schematics, diagrams and other system or component descriptions that define the configuration and design characteristics of the product,
- Material parts lists and information on the properties of the materials to be used,
- Information on processes, procedures, manufacturing techniques as well as instructions on installations or product processing, specifications for procurement and storage,
- Test instructions including necessary test steps and, if applicable, acceptable results and tolerances including associated test devices,
- Compliance documents or in the form of test results, assessments and/or calculations.

In addition to the manufacturing and maintenance documents, the design output also includes operating procedures. They serve the purpose of providing users with information on intended use, safety procedures and product preservation. Typical operating specifications are, for example, operating manuals and instructions.

In accordance with Sect. 8.3.5 e), special attention should be paid to critical items and key characteristics. For these, design outputs must be subject to special control and risk assessment. For example, stricter specifications for inspections or stricter tolerances for manufacturing or maintenance may be required. Warnings or notes in operating manuals may also be possible.

According to Sect. 8.3.5 f.), design outputs must be released by authorised persons before publication. In addition to an internal authorised person, this may also be the customer or an authority. This requirement is also contained in the EN section on the preparation and updating of documented information (7.5.2 c).

8.3.6 Design and Development Changes

Products and services change during their life cycle due to general improvements and innovations, design modifications, customer requests or repairs. The necessary design changes must be worked out in a structured and comprehensible way. Regarding the basic requirements for the process, design changes differ only slightly from new developments. In any case, the change process is inseparably linked to the configuration management in accordance with Sect. 8.1.2 of the standard.

Irrespective of the type and scope of the change, the associated design process is normally divided into the following components:

- Initiation and commissioning,
- Assessment (in particular evaluation of impact and risk analysis),
- Approval/release,
- Implementation, monitoring and documentation.

The first step is to describe the need for change. This is usually done by means of change request.[17] In this, the initiator of the change specifies the product affected and justifies the change or the problem (Why is the change necessary)?

In the course of the initiation, the advantages, risks and technical effects should be mentioned, as well as an initial estimate of the required time and costs.[18] A well-founded change request provides an appropriate basis for decision making. Depending on the type and scope of the change and the organisational structures, the change request is decided by the management, the head of department or the responsible design. In large companies, the task is often performed by a change board, which usually includes not only design staff, but representatives of all (potentially) involved departments. Customers and suppliers are also integrated in case of strongly networked supply cascades.

While small changes are often released at the same time as the assessment, more complex changes are released in a separate phase. For this purpose, the design department and other organisational areas (e.g. logistics, procurement, production) are requested to make a detailed assessment of the feasibility, effort and influence on the product or service. In this context, it must also be examined what effects the planned change will have on existing products that have already been delivered. At the same time, a process must be defined that regulates how customers are informed about design changes before they are implemented, if the planned change has an impact on their product or service.

The planned change must then be worked out to such an extent that physical implementation can begin after its approval. For example, calculations and simulations must be carried out, technical effects described, verifications and compliance with any legal regulations demonstrated as defined. If necessary, validations are also necessary. At the end of the evaluation, the change is approved and released by the operational decision-maker. If the persons responsible finally release the change, this is the starting point for the implementation and its monitoring. The specifications such as design drawings, software programs, parts lists and diagrams are then created.

At the end of a design phase (decision gates), results are verified by the change committee. In accordance with Sect. 8.3.4 h) of the standard, these meetings also include a formal, i.e. comprehensible, release for the transition to the next design phase. Intermediate approvals may also be necessary, in order to prevent the change project from running out of control, e.g. if the project is in a critical status.

The documentation requirements are based on the type and scope of the design changes. The minimum requirements are set out in the list a) to d) of EN Sect. 8.3.6. Accordingly, documented information (i.e. specifications or evidence) must be produced at least about the changes themselves, the associated evaluations, the approval conditions and measures against undesirable impacts.

---

[17]In the design phase, changes are usually controlled by a *change request* (CR), change *proposal* (MP) or *change* notification (MOD).

[18]cf. ISO 10007 (2004), Sect. 5.4.3.

For reasons of project control, traceability and risk considerations, the degree of formalisation should be higher the more complex, critical and cost-intensive the changes. Here the requirements for changes do not differ significantly from those for new designs.

8.4   Control of Externally Provided Processes, Products and Services

No company can rely solely on their own resources to provide products and services. Due to the constantly increasing specialisation, partial outsourcing of the value chain has become more and more important over the years. With the last revision in 2016–2018, the standard was increasingly adapted to this development, with outsourced processes and purchased services gaining in importance. As a result, linguistic adjustments had to be made at certain points. Thus, the term procurement was replaced by external provision. In addition, the concept of external providers was introduced. This includes all external suppliers of products and services, e.g. suppliers, subcontractors or service providers for outsourced processes or outlocated work, affiliated companies such as subsidiaries, sister companies or parent companies (outside the scope of certification). In the further course, the terms supplier and subcontractor will continue to be used here.

The standard requires a systematic selection of external providers (8.4.1), appropriate communication of the requirements relevant to the contract (8.4.3) as well as situational control and monitoring of all external providers (8.4.2) (Fig. 8.8). This does not mean that only Tier 1 suppliers are subject to monitoring

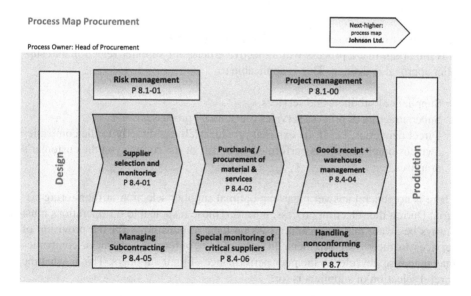

Fig. 8.8 Example of a process map procurement

by the company. Depending on the criticality of the products or services supplied, monitoring at lower levels of the supply cascade may also be necessary. In addition to criticality, the ability of the supply chain to pass on customer requirements downwards is also decisive.

Subsection 8.4 must be checked by the auditor during *each* EN certification audit.

8.4.1   General

At the beginning of a procurement there is the determination of the product to be procured or the service used. This must be exactly defined. The most important information is usually:

- Product and service characteristics,
- Price and conditions of delivery,
- Delivery times.

Further criteria for supplier selection and evaluation are also

- the criticality of the product or service,
- the qualification of the supplier (e.g. product portfolio, references, certificates, approvals),
- own experience with the supplier,
- the type of cooperation (e.g. subcontracting), criticality of the supplier/dependency for own value creation,
- sustainability of the supplier's market presence.

Requirements for the Selection of External Providers
It is important that a process with objective criteria for supplier selection and supplier approval is defined. This is applicable to

- Suppliers of products and services,
- Subcontractors of process services (outsourced processes),
- Direct deliveries, i.e. if the external provider delivers directly to the companies own customer or work is performed on behalf of the company at the customer's premises.

There is no general answer regarding optimal supplier selection or monitoring criteria. During the selection process, however, those responsible for operations must always be aware that they cannot pass on their responsibility for the provision of services to their own suppliers. In case of nonconforming products, companies may therefore not refer to the poor performance of their suppliers. This is what the careful selection of suppliers is for.

The company must determine the individual risk of each supplier (e.g. dependencies, criticality for its own performance) and align supplier management accordingly. Under certain circumstances, risk minimisation measures (e.g. alternative suppliers, design changes, more precise incoming goods inspection) must be taken. In the case of more complex or risky orders (financially, technically, time), a new order-specific risk assessment may have to be carried out before the order is placed.

The company must ensure control over the entire supply cascade. If the own supplier also outsources the contracted service, all original customer requirements must still be fulfilled. The company can ensure this control either by excluding delivery cascades in the contract with the external provider. Alternatively, the company requires a right for final release in case that additional subcontractors are involved.

On the other hand, customers occasionally either demand a right to release new sources of supply or they wish to have certain suppliers commissioned. This is particularly widespread in special processes (e.g. electroplating, welding or painting).

The company must then take these customer requirements into account in its procurement processes.

Possible Definition for Critical or A- Suppliers
- tend to be highly relevant for the fulfilment of the customer order (project-critical parts)
- the services are critical in terms of price, deadlines and quality
- parts are technically critical for product
- highly relevant A-supplier with a high order volume (high relevance for the company)
- poor delivery performance

No matter how the company selects suppliers, the procedure must always be transparent. It is important that the decisions and measures of those involved remain traceable throughout the entire procurement process. It must be evident from the records that decisions on the selection, monitoring and evaluation of suppliers are made on the basis of comprehensible evaluation criteria. If, for whatever reason, this cannot be ensured selectively (e.g. due to missing data from purchasing or incoming goods, no certificates or audits), a file note must be created so that the decision basis remains verifiable. The same applies if deviations from the objectively expected decision-making behaviour occur during supplier selection (e.g. if a supplier is selected despite poor experience and alternative suppliers).

Only suppliers of products and services that have no direct influence on the own value chain may remain excluded from systematic selection and monitoring. In the case of a production plant, these are e.g. office supplies, notebooks, printers and other small electronic office devices, catering, presents, gardening activities, cleaning of buildings.

It is important to determine the risks when selecting suppliers (e.g. experience with requested services, single-source, delivery time). For this purpose, evidence of risk assessment and any measures must be available in the certification audit. The type and scope of these should be based on the risk or criticality of the supplier or the product or service to be delivered.

Specific Requirements for the Selection and Monitoring of External Suppliers
The enumeration a)–e) in standard Sect. 8.4.1 formulates detailed requirements for the selection and monitoring of external suppliers.

(a) A process for supplier selection and release must be defined. This must be available in written form in accordance with Sect. 4.4.2. It must regulate which requirements external providers have to meet in order to be approved as a supplier (usually clustered according to A, B, C suppliers). Once the quality maturity of the supplier has been checked and given, it may be released. It must be defined here which inspections and evaluation criteria are the basis for this process. In addition, monitoring criteria as well as the scope of approval and the approval status must be defined. Suppliers who are not eligible for approval must be blocked. At the same time, it is also necessary to determine the conditions under which a blockage may occur and be removed.

(b) Each company must maintain a supplier register. This is usually done automatically in purchasing via the ERP system. Herein, the scope of the supplier approval must be documented, i.e. any restrictions on the type of product (e.g. not approved for flying material) or certain activities (e.g. only milling, no electroplating). Furthermore, the release status must also be defined (e.g. released, conditionally approved or blocked).

(c) Suppliers must not only be checked at the beginning of the business relationship but must also be constantly monitored. A distinction is made between
   – an evaluation of the deliveries. At a minimum, the punctual delivery performance (OTD) and product conformity must be measured. Other parameters can also be included in the assessment, such as the results of reaction times to enquiries or the quality of proposals.
   – a periodic evaluation of quality maturity level (independent of individual orders), e.g. based on ISO/EN certificates, regulatory approvals, own supplier audits or (documented) discussions with internal departments. The evaluation effort should be based on one's own experience and on the importance of the supplier (see 8.4.2 c). For this purpose, clusters can be formed which reflect turnover, criticality and qualification. Then, however, it must be checked periodically whether originally less important suppliers have gained in importance and are to be monitored more intensively. In any case,

the more important the supplier or its service, the more information must be gathered. In this respect, supplier audits, risk analyses and/or material tests may become necessary for important or critical suppliers.

On the basis of this doubled (order-independent and order-related) monitoring, it must be checked approximately every one to two years whether the supplier still meets the expectations. If a supplier continues to meet the quality requirements after periodic inspection, the approval status may be extended. If necessary, the status or the scope of approval must be adjusted. The NOTE in the standard explicitly points out that the type and scope of monitoring can be made dependent on the suppliers ISO or EN certifications or approvals by the aviation authorities.

(d) If order-related complaints are identified in the delivered products or in the supplier's services, the supplier must be urged to correct or rectify these problems. In case of serious deficiencies, the implementation of related measures shall be monitored. If the supplier is unable to provide adequate quality or performance (perhaps despite the possibility of rectification), the supplier must be blocked from further orders. If the supplier cannot be blocked due to a monopoly position (single source), at least stricter incoming goods inspections must be instructed and the responsible employees in design and production must be made aware of this supplier risk. If necessary, the customer must also be informed about the supplier's low performance.

(e) Internal requirements must be defined for document control of the external provider (specifications and records). For this purpose, e.g. specifications for verification, procedures for release or handling of incorrect documents, for data access and backup as well as for storage periods and storage conditions are to be defined. In operational practice, this EN requirement can be covered, e.g. by regulations in the quality assurance agreement with the supplier.

Focus During the Certification Audit

In a certification audit, the supplier selection and monitoring process is checked by using two to three suppliers as examples. These must indicate the basis on which the supplier was selected. For this, it must be recognisable who released the supplier at what point in time based on which information. An additional part of the audit is always the risk analysis. Furthermore, the measurement of the on-time delivery and the product conformity is usually verified on a random basis. Part of the sample is often a supplier with poor delivery performance. For this, it must then be demonstrated that appropriate measures have been initiated to improve its service provision.

In addition, the supplier register must usually be presented in the audit, including approval status, approval period and scope of approval. However, this can be an electronic list in an Excel spreadsheet or as an overview in the ERP system.

8.4.2   Type and Extent of Control

The company as the client must ensure that purchased materials, products and services are of a quality that allows to take full responsibility for them.[19] The company must therefore not solely rely on quality promises made by its suppliers. In this respect, the delivered products and services as well as outsourced process services of external suppliers are to be monitored.

Outsourced processes must be carried out under the QM system of the company (see 8.4.2 a). The type and scope of monitoring are determined by the individual circumstances and capabilities of the supplier. The type of products and services supplied also plays a role. Supplier monitoring can range from random inspections to ongoing process support (e.g. cabin modifications, complex engineering services) and detailed acceptance tests.

According to 8.4.2 c), the following aspects mainly determine the scope of the suppliers monitoring during or at the end of a service provision:

1. the type of the products or the service package to be performed. The scope of monitoring depends on whether the service provision is characterised by a stable, simple, possibly repetitive value creation process (e.g. simple products or series processing) or whether it is a complex, moderately transparent work package (e.g. sophisticated products as part of a one-off production, zero or small series processing). The easier it is to monitor the provision of services and the easier the processes are to carry out, the sooner the level of monitoring can be reduced and kept low. Furthermore, the importance of the delivered service for the own product or the own value chain plays a role for the intensity of control. For example, equipment or outsourced auxiliary work are subject to less intensive surveillance than the monitoring of outsourced special processes, engineering services or the purchase of critical items. A high level of control intensity is also expected for raw materials used in critical items. The handling of such products should be regulated by operational risk management, usually through regular validation of raw materials.[20]
2. the control and management measures that the supplier shows to his client. This includes, above all, appropriate communication with the client and the monitoring system of the supplier's own value chain. Finally, the experience of the external provider with comparable work or the experience with the technologies and processes used also matters for the scope of monitoring.
3. Closely related to point 2) is the requirement to make supplier control dependent on the results of supplier monitoring (OTD and conformity measurement).

If a supplier does not have the desired qualification, the company may have to provide support in building up know-how. Therefore, the client temporarily supports

---

[19]This shall also apply for reasons of product liability.
[20]cf. IAQG (2019).

the work of his subcontractor on-site with his own production staff for the purpose of on-the-job training.

During the execution of the work or at the latest with its acceptance, the results of the outsourced services are checked against the agreed requirements. The nature and extent of this verification depends on the risk in case services that do not meet the requirements are transferred to the operational value chain. External services or processes do not differ from purchased products. The company must be aware that it is not permissible to pass on the responsibility for verification to the customer.

The most common verification activity on products is the simple incoming goods inspection, in which at least a visual inspection of the packaging and the product (e.g. surface damage, deformation, corrosion, etc.) and a completeness check are carried out. An analogous simple check is also possible for services or documents. Regardless of whether product or service—in both cases there is always a comparison between order, delivery note, and service provided.

In addition to the incoming goods inspection, many products and services are subject to further verification activities:

- For standard parts, raw materials and consumables, but also certain services, it is common, especially for large quantities, to only inspect products on a random basis. In this case, statistically recognised procedures must be applied (cf. Sect. 8.5.1 c) 2).
- For unreliable or critical suppliers or critical services, supplementary inspections or an increased inspection severity are usually applied in addition to the regular incoming goods inspection. Under certain circumstances, extended documentation requirements in the area of material or product certification may also have to be met. In the case of critical raw materials, detailed product or material testing is necessary.
- Where applicable, verification activities must also explicitly focus on the identification of counterfeit parts. For methods and measures see also Sect. 8.1.4.
- For parts of high-value or complex services (e.g. drawing parts), the product acceptance must not only be carried out against the specification or order requirements, but may also be supplemented by photos, data queries and test instructions. In these cases, characteristics of product performance are being tested (e.g. dimensions, functions, degree of performance). In addition to these, there must be test specifications that provide information on what exactly is to be tested and which tolerances are permissible (e.g. using test instructions or test procedures). The verification may have to correspond to the scope of a first article inspection.
- If the company purchases drawing parts or process services, it does not transfer the responsibility for the appropriate provision of services to its subcontractor. For this reason, monitoring and testing activities shall be carried out to an extent necessary to ensure that outsourced processes and functions are performed as required (see also 8.4.2 c). Occasionally this is associated with verification activities even during service provision of the supplier.

If the company uses test reports from the executing supplier as part of testing or monitoring activities, there must be a process for evaluating this test data. The focus must be on checking the test conditions against the test requirements specified by the company or its customer. Verification can be carried out, for example, by auditing or by submitting validation reports.

Only if the delivered product or service corresponds to the purchase order, i.e. if there are no defects or discrepancies, it may be integrated in the own material flow. Until then, the goods shall always be considered as blocked. This concludes the control of externally procured services.

If the delivered goods are not in the required condition, they must be recognisable as such and kept in a restricted area until clarification is obtained. This is necessary in order to minimise the risk of unintentional feeding into the operational material flow. In the event of obvious transport damage, the delivery should only be accepted subject to reservation and this should be noted on the delivery note.

If procured products are planned to be processed before the required verifications have been completed, special attention must be paid to traceability. In addition to a special release in goods receipt, the affected products must be labelled, and the missing completion of the verifications noted in the product accompanying documentation.

If verifications, which are actually the responsibility of the company itself, are delegated to suppliers, the nature and scope of these activities shall be specified when placing the order. No grey areas may arise here. An overview shall be kept of all outsourced verifications in order to reduce the risk of loss of control. When assigning verifications to a supplier, it should be noted that the activity itself can be delegated, but not the underlying responsibility. In this case, the company must regularly carry out its own control and monitoring activities at the supplier's premises in order to verify its reliability.

For reasons of traceability and for supplier evaluation, records on the verifications of procured products or services must always be kept.

8.4.3 Information for External Providers

The most important criterion for purchased products and services is their compliance with the procurement requisitions. For this reason, the procurement data must contain technical details and key characteristics of the goods or service to be performed. Typically, this is done using the supplier's catalogue description and order number. Specifications are used for drawing parts or other non-standard products and services. The product or service description must be as precise as possible. Finally, these documents are enclosed with the order, because they are an essential part of the contract between the company and the supplier. Particular attention is paid to the procurement details in the purchase order (specification, contract, quality assurance agreement, etc.), as these clearly define the product or service to be procured from the supplier.

Possible Information When Using Subcontractors

- What should the subcontractor do (specification of the service package to be provided)?
- How is the schedule structured (e.g. for delivery, milestones, material to be provided)?
- How does communication take place between the company and subcontractor (e.g. contact persons, reporting, notification of expected missed deadlines, cut, nonconforming products)?
- What documents are provided to the subcontractor (e.g. specification, drawings, schematics, CMM, IPC, information on standards to be met, templates)?
- What documentation does the subcontractor have to provide?
- Which test requirements and records are to be provided by the subcontractor and which verifications/validations are to be executed by the company?
- Which certificates should the subcontractor enclose after conducting the work (CofC, EASA Form 1, etc.)?
- How should the subcontractor's supervision be structured, and which milestones or inspections are necessary? This is particularly relevant for special processes (where the quality of the output is not immediately apparent, e.g. electroplating, welding)?
- Which contributions or support services are provided by the company (buyer furnished equipment (BFE), resources, transport, etc.)?

The highest priority is to describe the product or the service to be provided (Sect. 8.4.3 a) including associated inspections, tests and release activities (b and i).[21] This applies in particular to the context of critical items (h) and in case of applying statistical test methods (j). In addition, the purchase order must contain information on special requirements regarding manufacturing processes (e.g. for special processes), equipment as well as for shipping, storage and transport.

If services are procured externally, the qualification of the personnel employed by the supplier (c) can also play a role. Therefore, any requirements regarding personnel qualifications must be communicated to the subcontractor in writing (e.g. certificates or professional experience).

Another important aspect in the cooperation between the company and subcontractors is the coordination and control of order processing (d to g). This includes, for example, the definition of interfaces and responsibilities or the determination of the type, scope and frequency of communication. The outsourcing of design activities in particular requires a detailed procedural framework.

---

[21]EN 9120 does not include all requirements of Sect. 8.4.3 of the basic standard EN 9100 and therefore has a slightly different sequence.

Requirements for the supplier's QM system play an important role in the procurement of products and services. For example, there is a widespread obligation in the aerospace industry that the supplier must maintain a QM system that is certified in accordance with the EN 9100 series. In EN Sect. 8.4.3 further possibly necessary obligations of the supplier towards the customer are listed:

- Willingness to use prescribed sources of supply and to have changes to them approved
- Obligation to notify of nonconforming products, services or processes for their disposition
- Take measures to identify or prevent counterfeit parts,
- Applying for approval in case of product modifications or changes in production processes, locations and external suppliers
- Obtaining an approval from the company, if the supplier plans a further subcontracting (delivery cascade). In this case, the sub-suppliers must meet the same quality requirements as the Tier 1 supplier,
- Provision of test specifications and test equipment if necessary,
- Documentation requirements, in particular compliance with retention periods and archiving conditions. Due to often very long retention periods, it should be clarified how to deal with documented information in the event of termination of the business relationship, transfer of business or insolvency of the supplier.

In addition, when external services are purchased, access rights to the supplier's relevant production facilities must be guaranteed for the organisation, its customers and regulatory authorities.

Inaccuracies in the procurement details can cause misunderstandings within the contract performance and can thus lead to additional costs, rework, quality losses and delays in delivery. In this respect, especially in the case of more complex purchase contracts, it should be checked by the purchasing department or the operational requester whether sufficient order information has been specified. Otherwise further specifications, e.g. in accordance with EN Sect. 8.4.3 a)–m) must be included in the purchase order or contract.

In practice, some of the requirements of Sect. 8.4.3 are often specified using text modules in contracts, quality assurance agreements or passages in the general terms and conditions. For the general terms and conditions of business, however, the effectiveness or the obligation to comply is limited. In case of doubt, often only the commercial law is valid in the end. Quality assurance agreements should therefore be concluded at least for subcontractors.

In accordance with Sect. 8.4.3 m), the employees of the external provider must also be aware of their contribution to product conformity, product safety and ethical conduct. The company can fulfil this requirement, for example, by informing the supplier of its compliance, product safety and ethical behaviour (e.g. quality policy) when placing an order.

As part of the certification audit, it must be expected that the auditor will focus on the adequate product or service description in the proposal and contract.

The focus can be based on a comparison of the revision levels instructed to and confirmed by the supplier on one hand and the document actually used by him on the other. In particular, the fulfilment or consideration of the requirements k) - m) must be regularly demonstrated in certification audits.

8.5 Production and Service Provision

8.5.1 Control of Production and Service Provision

Section 8.5.1 summarises the essential requirements for systematically organised production and service provision. The creation of controlled conditions is in the foreground. The provision of services must therefore be planned and carried out in a structured manner as well as monitored and appropriately documented. This presupposes

- all necessary processes and activities are defined,
- specifications and instructions are available to an appropriate extent,
- the provision of services is monitored, and products and services are being tested before delivery,
- the necessary equipment is available and used properly,
- the staff is available and qualified to carry out the assigned work,
- the traceability of the production process including the sub-assemblies is ensured, and
- records are kept demonstrating that the service has been provided in accordance with the requirements.

In Sect. 8.5.1 several requirements are defined, some of which are specified in the further course of Subsect. 8.5 and can therefore be neglected at this point. The enumeration points of Sect. 8.5.1 are described below:[22]

(a) The product or service must be clearly described so that it is unambiguous to the staff what is to be done with what, how, and which characteristics and features the product or service should have at the end of processing. For this purpose, written specifications must exist, e.g. technical documents such as parts lists, drawings, photos, instructions, standards, etc. For production or processing, further specifications must exist beyond the documentation that directly describes the product or service. After all, the work must be structured, understandable and carried out in the right way. The necessary activities

---

[22]EN 9120 does not include some individual requirements here and therefore has a slightly different sequence. However, the standard for distributors requires additional measures against the consequences of obsolescence (materials, components, resources, products).

have to be carried out on the operational level systematically, repeatedly and traceably with the same quality. Process descriptions, instructions, guidelines, samples, templates or photos simplify individual work steps. Job cards, production orders, project files, forms and checklists etc. help to structure the work sequence and to evaluate or release the tasks. All these documents have in common that they give security to the personnel and thus support the correct and complete execution of the work. It is important that all instructions are available on site for the executing employee and not just "somewhere".

(b) In the course of service provision, monitoring and measuring must be carried out on the processes as well as on products and services. The necessary monitoring and measuring resources must be available for this purpose. It must also be ensured that the company controls these resources and uses them correctly. The focus of this requirement is not on a general provision and availability of these tools (cf. Sect. 7.1.5), but above all the correct use and, their availability for individual orders.

(c) The implementation of monitoring and measurement activities shall be defined in terms of type, scope and timing within the value chain. Details are defined in the standard's Subsect. 8.6 (Release of products and services) and Sect. 9.1.1 (Monitoring, measurement, analysis and evaluation—General). In addition to the specifications and QM documents, compliance documents, e.g. test results or activities performed (e.g. measurement results, calculations, etc.) must be provided.

In the context of monitoring and measuring activities, documented information (specifications and objective evidence) must generally exist for the following aspects:

– Acceptance or rejection criteria.
– At which point or at which production step tests are to be carried out.
– Requirements for the recording of test results. Forms and checklists can support structured processing and facilitate the recording of test results.
– Specifications regarding the measurement and test equipment to be used (e.g. mechanical or electronic measurement equipment, templates and measurement standards) and, if applicable, instructions for their use.

If tests are carried out based on random samples, it must be ensured that statistically recognised methods are used, e.g. in accordance with AQL based on ISO 2859. Moreover, the specific test situation must be suitable for a random sample. It is also important that the criticality of the parts or products concerned is considered when determining the sample size. In operational practice, it is not uncommon that the samples are not statistically valid due to inadequate or missing use of methods or that the risk associated with the product was not considered in the planning of the test.

(d) It must be ensured that the company maintains and uses the necessary order specific infrastructure (including resources/tools) to an appropriate extent and guarantees a suitable working environment. This is not a matter of general provision (cf. Sects. 7.1.3 and 7.1.4). At this point, the focus is on order or project specific suitability. This means for infrastructure, operating resources and equipment that they are currently approved and suitable for the task and are used under controlled conditions. This also includes that the respective location of the equipment is known and that the condition and completeness of the equipment are checked as part of regular inspections. If an equipment has been lost, the organisation shall apply a clear procedure. Records must be kept of controls and any losses. For machines requiring regular maintenance, maintenance schedules and specifications must be available, and records of the measures taken must be kept (see also Sect. 8.5.1.1).

(e) Personnel shall be authorised and competent to carry out the work assigned to them in a manner appropriate to the requirements and within a reasonable time. The standard provides detailed requirements regarding personnel competence and authority in Subsects. 7.2 and 5.3, so that a discussion is not required at this point.

(f) Special processes must be adequately monitored and validated. Section 8.1.5.2 is solely focussing to this issue; a discussion is therefore not required at this point.

(g) Measures shall be taken to prevent human error. These include, for example, double inspections, continuation training, lessons learned meetings and special instructions or notes on job cards for activities that are known for frequent human error. Above all, the critical processing of errors is important. To initially identify the reasons of human error, a short look at the Dirty Dozen is often very helpful (cf. Fig. 8.9).[23]

(h) Requirements for the release of products and services and their support after delivery are formulated in detail in Subsects. 8.6 and 8.5.5. This means that both standard specifications can be neglected here.

(i) Operational activities require so-called Good Workmanship, which should also be a matter of course for a less experienced (but actively thinking) employee. If requirements go beyond this and the clearest practical method of implementation is unclear, then more detailed instructions must be formulated. However, this is already required in enumeration point a) of this section and therefore need not be considered here again.

(j) Responsibilities for order processing during the production process must be specified. In repetitive manufacturing, this is often defined by an employee from work planning. In other companies, the few employees of production control are responsible for all orders. Details are regulated by the responsible foremen, team or shift leader for their respective workstations or production areas.

---

[23]The Dirty-Dozen concept is a list of the twelve most common reasons for human error. Gordon Dupont, an employee of the Canadian aviation authority, evaluated the most common human errors in aviation accidents at the end of the 1970s and combined them into a conceptual framework. If these twelve types of errors can be minimised or brought under control, a very high percentage of everyday incidents and accidents can be avoided.

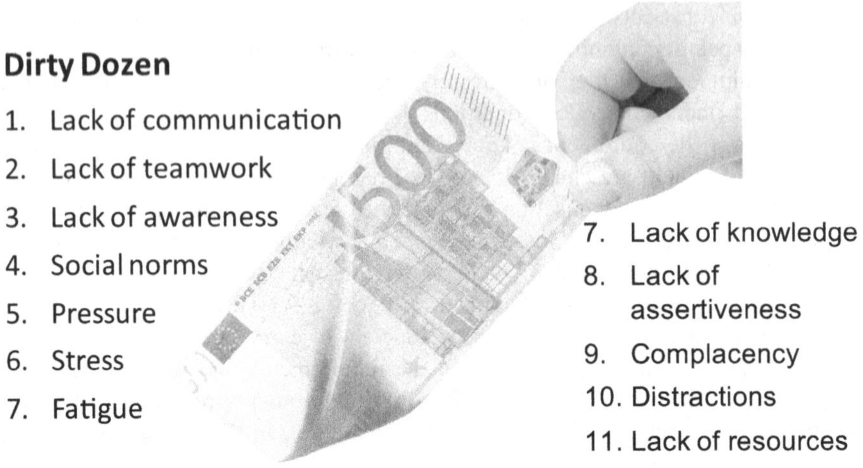

Dirty Dozen

1. Lack of communication
2. Lack of teamwork
3. Lack of awareness
4. Social norms
5. Pressure
6. Stress
7. Fatigue

7. Lack of knowledge
8. Lack of assertiveness
9. Complacency
10. Distractions
11. Lack of resources

Fig. 8.9 Dirty Dozen—the 12 most common reasons human errors. (see Sect. 8.5.1 g)

(k) The process design must consider the specific requirements of any critical items or key characteristics. This can be done, for example, by additional or more intensive testing/verifications (instruction for double inspections). This can be implemented in the manufacturing documents and in work instructions, by reduction of tolerances in technical documentation or by a special note "Critical component" in job cards. The design organisation, not the manufacturing employees, is responsible for identifying critical items and key characteristics. If a company operates exclusively as a manufacturer (Build-to-Print), critical items and key characteristics must be communicated by the responsible design organisation.[24]

(l) Methods and tools must be determined to measure variable data. This includes quantities that can have any value within a range (e.g. viscosities or processing temperatures). Tools for the measurement of these variable data and its tolerances must be clearly defined. An occasional problem that occurs is imprecise specifications, e.g. the calibration of measuring instruments at room temperature (but the question is: What is meant by "room temperature"?).

(m) Inspection of work steps usually only takes place after the completion of activities. However, the production process can conflict with this procedure if tests at a later stage are no longer possible (for example due to lack of accessibility). In such cases, intermediate inspections must be identified and considered in the production specifications.

(n) The requirement for documentation of all work steps is intended to ensure that the product remains identifiable during value chain and that the product

---

[24]See IAQG (2019), p. 9.

status, production progress and delivery status can be identified at all times. This requirement can be neglected here, since Sect. 8.5.2 of the standard deals with this topic in detail.

(o) At this point it is not necessary to take a closer look at this enumeration point dealing with prevention, detection and removal of foreign objects. This requirement is repeated in more detailed in Subsect. 8.5.4 b).

(p) In addition to the product itself and its components, auxiliary and operating materials must also be controlled and monitored. However, this must only be ensured if the substances have an influence on product conformity. For example, when cleaning with distilled water, a certain "purity" may be required. For galvanic product processing, the bath liquids must be checked continuously (special processes).

(q) The transfer of a service to the next stage of the value chain is only permissible after completion and execution of all planned measuring and testing activities. However, this requirement can be contradicting to day-to-day business. (e.g. on-site completion in/at the aircraft or system, deadline pressure). In this case, the product and its status must be identified and tracked so that the missing work can be traced and carried out at a later stage.

Not all requirements can (equally) be implemented in every company. For example, an organisation whose engineers verify design services on a computer or, under certain circumstances a company concentrating on milling, may waive the requirements f.) regarding special processes because these are not applicable to them. Companies (e.g. many service providers) that do not deliver in the classical sense can skip enumeration l. and p. Each company has to determine individually which requirements of Sect. 8.5.1 of the standard applies to it and which does not apply. In the certification audit, however, the auditor has to be convinced of this.

8.5.1.1 Control of Production Equipment, Tools and Software Programs

Since operational value creation can hardly be executed without extensive use of resources, these devices, including associated software, must be subject to controlled use and systematic monitoring. The aim is to control the equipment to such an extent that defects causing rejects, rework or defective products are reduced to a minimum. In addition, the operating costs of the devices are to be minimised through their structured control and monitoring and the longest possible service life is to be ensured. Validation and maintenance are the main instruments for achieving these objectives.

Equipment validation is necessary before the device is used for the first time and after maintenance, software release change, tool change, etc. At the same time, the first machined part must be checked to ensure that the production result corresponds to the specifications (dimensions, tolerances). Changes to the machine manufacturing parameters that influence the production output must be approved by an authorised person (e.g. the production or shift supervisor, see Sect. 8.5.6) after equipment validation.

If equipment requires regular maintenance, a maintenance plan (e.g. on-condition or fixed maintenance intervals) is to be defined[25]. Maintenance plans with clear specifications must therefore be available. Usually, the recommendations in the manufacturer's maintenance or operating manuals are used.

Records for validation or maintenance must be kept and contain information on the measure (time, parameters), the reason (e.g. first use, maintenance, release change), the person performing it and, if applicable, the underlying documentation (revision status).

The company may be required to maintain a register that provides an overview of the equipment used to enable control and tracking of maintenance activities. The minimum information of such a register should be based on the standard specifications for measuring and monitoring equipment (cf. Sect. 7.1.5).

When storing operating materials, it must be ensured that tools and equipment do not deteriorate due to unfavourable environmental conditions. For example, condensation or humidity can lead to corrosion of tools, devices or templates susceptible to rust. Dust, sunlight or extreme temperatures can also limit the usability of equipment; theft would even result in loss of equipment. In order to protect it over the long term, devices should therefore not be stored outdoors, they should be oiled as required, packed dust-proof or airtight and, if necessary, checked periodically for their condition. In this respect, it is important to define storage requirements for the equipment or to take it from the manufacturer's equipment manual. Compliance with the storage specifications must be monitored in day-to-day operations. In addition, the storage location must be determined in order to find operating resources again, even after years of non-use.[26]

Many companies have operating resources which are owned by the customer and for which the requirements of Sect. 8.5.3 must be applied.

8.5.1.2 Validation and Control of Special Processes[27]

Requirements for special processes are defined in this section of the standard. These are process steps in the provision of services, where the output cannot be directly tested without destroying the product itself or where carrying out a single test results in an economically unjustifiable effort. Without special test measures, errors during the creation of value would only be recognisable after the product is delivered and put into use. Typical examples of special processes are welding processes, heat treatments, electroplating, painting, gluing, sealing, coating and pressing.

Since product verifications are not directly possible for special processes, the validation of the associated processes (usually manufacturing processes) is used

---

[25]Many modern machines are self-maintaining. Here it is sufficient to document only those maintenance and repair events which are carried out beyond the mechanical self-maintenance.

[26]See also Sect. 8.5.4.

[27]This section is not applicable to EN 9120 for distributors.

as an alternative. Product testing is thus ensured indirectly through the evaluation of process parameters. Such influencing variables can be, for example, processing temperatures, humidity, stresses, mixing ratios or viscosities. Associated validation criteria for process quality are then randomised e.g. tensile force or material samples, hardness or bending tests or colour evaluations. For this purpose, acceptance criteria or tolerances must have been defined beforehand and be available as documented information.

An EN-certified company has to determine methods (tests, samples, tests) of how the validation is to be carried out and determine process parameters (voltages, temperatures etc.) of how the processing conformity is monitored after the validation, i.e. during operation. Only when the test results of the process validation are within the specified tolerances production can be released. It is also necessary to check whether additional requirements regarding personnel qualifications or the equipment are necessary in order to ensure process conformity (e.g. welding certificate or ovens with certain cooling properties).

For this reason, the company must explicitly specify (Sect. 8.5.1.2 b) after which period or events a new validation (re-validation) is required and which test conditions have to be evaluated.

Furthermore, in order to provide adequate evidence, it must be determined, how and which parameters are to be recorded in order to demonstrate the conformity of the special processes and the processed products.

If special processes are outsourced, the supplier monitoring process must meet the particular requirements of the subcontracted activities. Thus, the company has to monitor outsourced special processes and can therefore not rely on ignorance regarding the special activities executed by the supplier. In the certification audit at least a plausibility check of the process flow and the validations applied is expected, since in most cases no deep technical process knowledge is required for plausibility checks on special processes. In addition, the monitoring and measuring devices used and the methods for recording the results must be audited at the supplier.[28] If there is not enough experience within the company for such an assessment, an external expert should be consulted.

8.5.1.3 Verification of the Production Process[29]

Verification activities must be established as part of the operational production process to ensure that the production process meets all requirements as defined or for the intended use. Measures are given as examples in the NOTE of the standard's section. The verification of the production process includes all activities that serve to check the performance of the production process accordingly.

The First Article Inspection (FAI) is in the focus of Sect. 8.5.1.3 of the standard. At a FAI, a representative unit of the corresponding product is systematically put through its paces at the start of a new production run. The objective is

---

[28]See EN 9101:2018, Sect. 4.1.2.5 c.

[29]This section is not applicable to EN 9120 for distributors.

to provide evidence that the product can be manufactured according to the design data. The FAI proves that the main risks that may arise in connection with a ramp-up or after a change in the production process have been systematically examined.[30]

Therefore, it is checked, if the

- documentation,
- resources, tools and equipment,
- materials

are suitable to manufacture the product according to the specifications and requirements. In detail, the FAI requires individual tests to be carried out on all product requirements.

For this purpose, the following shall be examined:

- compliance with product characteristics such as 4 F (form, fit, function, fatigue), dimensions (length, width, height, gap dimensions, etc.), weight, touch and feel (e.g. grain) or risk of injury,
- configuration status,
- identification of the product and its components according to the specification,
- compliance with the functional test requirements,
- full execution of approved quality assurance planning,
- completeness, precision, traceability, conformity of production documentation,
- accuracy, completeness, traceability and correctness of the *compliance* documents (test specifications, test reports, certificates), and
- possibly also commercial aspects, such as compliance with the contract conditions (price, delivery date, completeness, shipping conditions) when using suppliers.

In addition, all manufacturing documents, test plans and test specifications as well as the related process instructions (including the revision status) must be listed in the FAI-Report (FAIR). The same applies to the relevant operating and testing equipment.

Records must be kept of the results of first article tests.

It should be noted that it may have been agreed in the contract with the customer to involve him in the execution of the FAI or to transmit the FAI results to him. A partial or complete repetition of the FAI becomes necessary if design changes have been made to the product with potential influence on the product characteristics or functions. A new FAI is also required for changes to the production process (modification of process parameters) or for a significant revision of production or test documents.

---

[30]Further information on the requirements of a FAI can be found in EN 9102 Aerospace Series - First article inspection requirements.

A product may only be released for series production if all nonconformities identified in a FAI have been comprehensibly corrected.

8.5.2   Identification and Traceability

Identification

A company must be able to guarantee reliable identification of products and services, including the current processing status or degree of completion, at all times during the value chain.

In order to identify the status of service provision, products and services must be identified during service provision. For services, the processing status must be documented using associated documents or a separate accompanying documentation. For products, identification is usually ensured during processing and storage by means of an accompanying document, e.g. with the order or a marking already affixed by the supplier. The accompanying material documentation is not removed and archived until the material, part or product has been incorporated or installed in another component (next higher assembly) and, if necessary, replaced by a certificate (e.g. CofC or EASA Form 1) before delivery.

At the same time, it is always necessary to mark the product yourself, e.g. by means of embossing, etching, tags or stickers. Only the smallest parts (screws, nuts, capacitors, etc.) are sometimes not marked at all or only partially. In this case however, it becomes even more important to ensure that the accompanying documentation is complete and correct, e.g. lot or batch number on the associated bags or containers.

Acceptance Authority Media

Activities on the product must be clearly and comprehensibly confirmed by the person conducting them. For smaller companies, this is usually done by means of a signature or abbreviation (incl. date) on the job card or work order. In this case, a reference list may have to be available so that an assignment to the relevant employees is possible.

Larger companies often work with personal stamps or electronic identification cards. In this case it must be ensured that such aids are "controlled". This means that their issue and collection or loss must be documented with the recipient's name, date and signature. This applies analogously to electronic release cards.

Traceability

Due to an increasing division of labour and an increasing internationalisation, companies are forced to source out more and more parts of their product and service spectrum to external sources. For these purchases, they must bear quality and liability responsibility. This is only possible if companies can trace and prove when they bought which products from whom or whose services they used and how they further processed these during their own value creation. Therefore, companies must ensure traceability. This requirement is also an important requirement in aviation legislation.[31]

Although the standard requires traceability for *both* products *and* services, there are some special requirements for physical products.

Traceability of Products
For materials management, traceability means that products and parts, materials and the substances they contain can be traced back from the source of manufacture or origin to their use or installation, scrapping or transfer of ownership at any time through the documentation. Here, operational material control is in contrast to strict requirements because all product movements and processing must be documented. Not only incoming and outgoing goods or storage must be traceable, but also (batch) separation, packaging and preservation processes. The same also applies to the provision of services by sub-suppliers. In daily practice, this regularly confronts companies with real challenges because it also means that, for example:[32]

- traceability back to serial number or badge/lot number is to be ensured—even if the part or material is integrated in a component or subassembly.
- the traceability of the entire product creation process is to be ensured (e.g. also conservation processes by suppliers) and differences between the target and the actual state of the product must be shown.
- all products manufactured from a raw material or production batch must be traceable from the source of purchase to delivery or scrapping.[33]

It must therefore be possible to provide information on the product status and its entire history at any time.

The distributor standard EN 9120 goes a step further here. It explicitly requires product labelling and traceability from goods receipt to delivery, including separation, storage, packaging and preservation. This also applies if one of these activities has been outsourced. When delivering separated products, the following documented information must be kept by certified EN 9120 distributors:

- proportion of delivered quantity of batch, lot,
- order number(s),
- the customer name.

[31]See EASA Implementing Rule Continuing Airworthiness EASA Part 145.A.42 (a) (5) und Guidance Material 21A.139 (b) (1).
[32]cf. Hinsch (2019), p. 260.
[33]This means, for example: If 50 m of hydraulic lines were procured which were installed by the company in three different aircraft systems, it must be known at all times in which aircraft and where the installation took place.

Whether EN 9100 or EN 9120, an EN-compliant traceability presupposes a monitored and traceable document flow, thus, a formal process to control product documentation. This is the only method to ensure that responsibilities, product history and sources of error can be clearly identified and addressed in the long term.

The scope of traceability is based on the company's own operational specifications or customer requirements. Full traceability—even in aviation—is only necessary for critical parts or on customer request. For parts that do not have to be tracked continuously, it is permissible to ensure traceability starting from the sub-assembly or assembly level. In operational practice, individual parts such as capacitors, resistors or nuts and bolts are therefore occasionally excluded from complete traceability.

If there are no explicit instructions regarding the scope of traceability in official requirements or in the contract, a specification should be requested from the customer.

It must be considered that the more detailed traceability is, the greater the effort, however in the event of product defects, this allows a the more precise determination of the affected quantity and thus reduces recalls or modification measures.

8.5.3   Property Belonging to Customers or External Providers

Material or immaterial property of customers, suppliers and partners are found in most companies and thus are a part of the daily practice. Most importantly third-party property plays a role regarding material provided by customers as well as for repairs, product modifications and on-site assignments in third-party premises. Intellectual customer property includes, for example, electronic and paper-related design documentation (drawings, analyses, parts lists, etc.) as well as production and maintenance documentation.

Dealing with third-party intellectual property also plays a special role in EN-certified companies, especially where their own performance is deeply integrated into the value chain of the customer, e.g. in design or in complex projects. Third-party, personal data can be significant when employing temporary staff.

Every company must define rules for dealing with third-party property. A fundamental requirement is the careful handling of third-party property. A condition and completeness check at the time of taking over the property shall also be part of a procedure. After the transfer, adequate protection against unauthorised access or use and against environmental conditions (weather, temperature, humidity, possibly also coffee cups at the workplace) must be ensured. If there is a risk of incorrect use or handling of third-party property, instructions must be made available to the company's own employees.

In addition, third-party property must be identifiable or recognisable as such, e.g. by means of tags or labels, appropriate accompanying documentation and separate storage. For documents, this can be ensured by a clear identification of the header or footer, alternative options are file names or clearly assigned IT folders.

Occasionally, these minimum requirements are specified in an agreement with the customer.

Records must be kept for incidents involving third-party property. Above all this serves to minimise the risk of subsequent liability risks and conflicts with the owner. The following documents or recordings may help for this:

- an inventory list of customer property, ideally including stock movements,
- documented takeover and exit controls,
- recordings of the associated customer communication,
- access requirements, confidentiality clauses,
- storage regulations, and
- data protection requirements.

A separate process description for dealing with customer property is not mandatory. In many companies, this topic is dealt with in the QM manual or distributed across various process descriptions or a separate instruction.

8.5.4 Preservation

Section 8.5.4 of the standard specifies requirements for the general handling of products and services within the company's responsibility. The corresponding specifications aim to avoid a deterioration of the condition. The focus is not entirely on the actual products or services, but also on the process results.

The idea behind it: if the process inputs and the internal processes conform with the requirements then so does the process output in the form of products and services.

Every company must demonstrate activities and measures aimed at preventing deterioration of process outputs (products, services, semi-finished products, results of sub-processes) and to maintain compliance with customer requirements. For products, the focus is on their handling and storage as well as on transport, packaging and shipping.

For provision without physical outputs the focus may be on loss of data, protection against unintentional changes or compliance with data protection requirements.

The fulfilment of the requirements dictates "good workmanship", which should be a matter of course even for a less experienced (but committed) employee. However typically only employees with specific product, service or process knowledge are aware of if these requirements have been met.

Examples of typical requirements for the preservation of products are listed below.

Handling and transport

- Order and cleanliness at the workplace should be a matter of course not only for an audit. This includes leaving the workplaces tidy at the end of an activity, the end of the day or at the change of shift. Food and beverages should not be placed directly at (production) workplaces as they pose a risk to the product condition. Documentation should always be organised, and unnecessary items should be protected in the event of non-use and returned to the intended storage location. These tips may sound natural however certification audits often include corresponding warnings or even findings on these issues.
- Production personnel should be trained to minimise product defects by preventing and detecting foreign objects (FOD). Special attention should be paid to the removal of contaminants and foreign objects that may damage or contaminate products or assemblies. Typical examples are not removed drilling chips, cable remnants as well as equipment or small parts left in the product (e.g. screws). As a precaution, appropriate tests (e.g. shake test, counts) should be indicated on the job cards.
- ESD[34]-sensitive parts can only be processed under professional protective measures, as they can be damaged or pre-damaged if handled improperly. In order to prevent any such deterioration during processing or handling as far as possible, ESD-protected areas must be established in which ESD mats or floors are laid out and the wearing of ESD clothing is mandatory. If necessary, the warehouse should also be set up for ESD. Controls of ESD tests performed are often part of certification audits.
- The production is to be carried out only at designated workplaces. This is not only because the necessary equipment may only be available there, but also because of any prescribed environmental conditions or protective devices. For example, small painting work can only be applied in the paint spray cabin. Soldering work may have to be carried out at ESD workstations rather than at one's own inadequately equipped workplace, even if this would be more convenient or easier.
- If necessary, clear transport and packaging specifications must be made and adhered to for sensitive parts and materials (e.g. protection by silver foil for parts sensitive to static charge, protection against vibrations from air-cushioned transport, refrigerated packaging, use of special containers or packaging to prevent corrosion or surface damage).

Storage

- Controlled storage conditions are intended to protect stored goods from deterioration. A prerequisite for this is the existence of and compliance with

---

[34]Electrostatic discharge.

product-related storage specifications. It is also necessary to be able to trace the storage conditions, the storage and retrieval processes and goods inspections via documented information.

- In order to ensure controlled stock movements, access to the warehouse should be restricted to authorised personnel. Otherwise, the risk of uncontrolled stock withdrawals that conflict with the need for traceability and product protection increases. The larger the production facility, the higher the corresponding risk.

- Environmental and storage conditions must not lead to a deterioration in the condition of the stored products and materials. For this reason, the environmental conditions must be monitored for many products. Thus, direct sunlight and excessive dust are to be prevented. In order to provide evidence of controlled stock management, temperature and humidity must usually be recorded via climate loggers. For this purpose, a procedure, including responsibilities, must be defined for when deviations from the acceptable range of the measured values are detected.

- Aerospace material shall be stored separately from non-aerospace material.[35] Unserviceable material must be labelled and stored spatially from serviceable material.[36] Nonconforming material or material with an unknown status shall be identified as such and placed in a quarantine warehouse to avoid the risk of accidental use.

- Materials with limited shelf life require storage time monitoring. For the specific handling of consumables, substances with a limited shelf life and its storage periods, the relevant data sheet shall be used. If no absolute best-before date (e.g. shelf life "BB until Jan. 2022") has been printed on the packaging, the corresponding data sheet for the substance must be used. In this case it is also advisable to use a shelf-life sticker on the packaging to indicate the delivery date and shelf life. The date of opening should also be documented on such a sticker. If it is ensured that the substances are consumed quickly, such strict storage time monitoring can be omitted (e.g. for soldering paste at soldering workstations). In almost every production facility, small storage containers of (decanted) liquids are found, which do not show what they contain and when they have been filled up.

- In order to minimise the risk of product or material damage during storage and retrieval, protective measures should be taken where appropriate. Their nature and extent depend on the value, hazardousness and sensitivity of the product or material as well as on the frequency of damage. Special storage regulations are usually taken from the material data sheets or manufacturer specifications. For example, surface damage must be minimised using separating protective

---

[35]This is mandatory for organisations with an EASA production or maintenance approval.

[36]This is mandatory for companies with EN 9110 and EN 9120 certification and EASA production or maintenance approval. However, "spatially separated" does not necessarily mean in another room, but physically separated. Therefore, another shelf also represents a spatial separation.

material; if necessary, additional gloves must be worn during handling. Numerous fibre composite materials must be stored flat and free of pressure load. Motors and gearboxes may have to be preserved before storage.

- Hazardous substances must be labelled with a warning and stored separately. An inventory and monitoring of hazardous substances must be ensured via a hazardous substance list or database. In addition, associated safety data sheets must be available as they contain information on the handling of the hazardous substances. These must be directly accessible to users (i.e. not in the office of the responsible material purchaser) so that immediate action can be taken in an emergency (e.g. eye rinsing with the appropriate substance). Where available, the Hazardous Substances Officer is responsible for the relevant administration. Otherwise, a responsible person must be nominated.

8.5.5   Post-Delivery Activities

The provision of services does not end with the delivery but extends to the period afterwards for complaints or guarantees.

Activities after delivery may also be triggered by the customer, by the legislator or by interested parties, for example due to:

- contract requirements (e.g. maintenance agreements, operational data evaluations),
- expectations of customers that are not fixed in the contract (including customer feedback),
- legal or official requirements (e.g. safety requirements, product monitoring, disposal).

Finally, operational requirements can also require extensive post-delivery activities, e.g. due to extended services to stand out from competitors or for the determination of long-term product and service quality. In the latter case, an essential element is the monitoring of products or services during the operating phase. The necessary information can, for example, be provided by transmitted customer data on product performance, customer complaints or fault analyses of maintained or repaired products. This performance monitoring can then be used, for example, to derive information on product reliability, the need for changes, improvement potential and risks, as well as information on future product design.

For such error analyses, a simple troubleshooting with low documentation efforts can be sufficient.

In the case of major or systematic product or performance deficiencies, however, more complex analyses of the root causes, patterns and the duration of errors must be carried out incl. detailed reporting.

The requirement 8.5.5 g) is derived from the EASA Part 21 J.Long-term updating, publication and distribution of the technical documentation for your own products must be ensured accordingly. This includes for example the publication of Service Bulletins (SBs), the technical preparation of Airworthiness Directives (ADs) and updating of maintenance specifications (CMM, IPC, etc.) or operating manuals. In this context, the requirements of EN Sect. 7.5 (documented information) and 7.3.5 (design outputs) shall be considered in addition to the EASA regulations.

The type and scope of post-delivery support are primarily oriented to the product or service. Some companies will hardly apply any of the requirements of this section, but there are also organisations that will have to take extensive action here. This section cannot be excluded completely from the scope of certification.[37]

A certification audit shall demonstrate, where applicable, a systematic approach to all applicable post-delivery activities through examples.

8.5.6   Control of Changes

The value chain is not a static structure but must be continuously adapted in everyday operations. The reasons for this can be found, for example, in the:

- modifications to products or services (troubleshooting, functional expansion),
- changes to the production process (sequence, insertion or removal of process steps, outsourcing),
- use of new machines, tools and new software, including release upgrades,
- change of performance parameters (soldering temperatures, machining speeds, e.g. during milling, heat treatments, cooling curves),
- use of new materials, operating or auxiliary materials (e.g. new adhesives, CFRP substitution).

From a standards perspective, it is important that such changes are planned and implemented in a structured manner. It shall be ensured that the changes are made before their implementation

- in type and scope,
- the influence is determined,
- necessary measures/activities derived from this, if necessary, and
- the entire procedure is adequately documented.

After implementation and before the processes are restarted, the effectiveness of the change must be determined by a new first article inspection (FAI) (Has the intended objective been achieved?). In addition, the maintenance of product or service conformity must be determined (Do the product/service properties still meet the requirements?). The nature and extent of such an evaluation shall be proportionate to the change. Examples of monitoring changes are:

---

[37]Cff. IAQG Resolution Log of Oct. 2019.

1. Example: If previously manual milling work is to be replaced by a new CNC machine, the change must be planned and evaluated. For this purpose, the effects and risks on the production process must be determined in advance, as well as downtimes due to the changeover and possible delays to promised delivery dates. After the installation of the machine, a test production run must be carried out and the milling quality evaluated with a first article inspection (FAI). It may also be necessary to check, if such a comprehensive change in the production process requires the customer's approval (based on the contract requirements).

2. Example: The introduction of an automated gluing machine instead of manual gluing, promises a constant continuity in the glued seam, uniform glue application and fast processing times. After the introduction of the automated gluing machine, a FAI must be executed to verify if the glued seam and gluing application comply with the specifications and if the processing time meets the expectations.

The change of a subcontractor is usually a change that requires monitoring in the sense of this EN 9100 section. Even the implementation of a production-relevant IT system is a change in the value chain, which therefore requires a structured action for the introduction with subsequent verification and release.

Changes to value-added processes may only be released once the planned result and the maintenance of the requirements have been comprehensibly determined by a person authorised to do so. The persons authorised to release process changes must be defined for this purpose. Usually this is a production manager or the head of design or the CEO. The authorised persons do not necessarily have to be defined by name but must be specified through a position (e.g. "the production manager").

The following documented information on the changes must be maintained:

• documents that describe the change of the product, service or the process procedure,
• evidence that the planned changes have been successfully implemented (e.g. measurement results, FAI protocols),
• reason for the change and
• release of the change by an authorised person.

During or after implementation of the change, the affected employees must be informed and, if necessary, trained.

8.6   Release of Products and Services

Products and services can only be beneficial to the customer if they meet the customer's requirements. If they don't, the customer's satisfaction drops. In order to ensure that all delivered products and services meet the defined specifications, tests (verifications) and releases are required.

   The EN 9100 requires a systematic procedure for monitoring and measuring the product properties. Inspections and evaluations are usually not just at the end of the value chain but appears across the entire process of service provision (cf. also 8.5.1 l). Intermediate inspections ensure that nonconformities are identified at an early stage and thus, rejects or rework are avoided in order to minimise costs and time loss. Intermediate inspections can take place at fixed inspection points in the value chain or during transfer between production sites or work steps, as well as at the end of service provision.

   During service provision, inspection points can also be set after document creation or before milestone meetings.

   Tests on products and services must be carried out based on planned regulations. The following requirements are usually to be defined in the context of product and service releases:

(a) Acceptance or rejection criteria.
(b) At which point or at which process step tests are to be carried out.
(c) Requirements for the recording of test results. Forms and checklists can support structured processing and facilitate the recording of test results, especially for complex services.
(d) Specifications regarding measuring and test equipment to be used (e.g. mechanical or electronic measuring instruments, templates and measurement standards, image samples) and, if applicable, instructions for their use.

In this context process descriptions, test specifications and instructions, technical implementation procedures, job cards or generally accepted test methods are essential documents for example.

   The monitoring and measurement of a product or service may involve visual and completeness checks, functional tests or measurements for example.

   In principle, the delivery of the product or service to the customer is only permitted after completion of all activities and their final internal release. However, this requirement may conflict with everyday business operations (e.g. on-site completion in/on the aircraft, meeting deadlines) or customer requests. In this case, the product and its status must be identified and tracked in order to be able to understand and carry out the remaining work. If the product is delivered at the customer's request before release, a customer confirmation must be obtained. This is preliminarily to be able to counteract any liability claims or customer complaints resulting from the early delivery.

   At the same time, customer approval increases awareness of the fact that the product may not be free of nonconformities and thus lowers the expectations. Before the unfinished product or service is delivered, a special internal release is necessary.

   As a note, useful key figures for evaluating process performance can be derived from the measurement and testing activities (see Chap. 9). Cumulative evaluations can be used to identify systematic nonconformities or deficiencies. Useful examples for cumulative evaluations are error or rejection rates as well as test effort (time or costs).

Appropriate evidence must be provided for release activities so that it can also be determined retrospectively that the product or service provided meet the defined requirements at every important point in the value chain. The test records must be available at the time of delivery.

8.7   Control of Nonconforming Outputs

Organisational processes are associated with an occasionally occurring improper provision of services. This can take place in the processes of the own value creation, but also in the value creation of suppliers or partners. If this results in defects or damage to the product or service, compliance with customer requirements may no longer be ensured.

Normally, nonconformities are identified via the following mechanisms:

- in the context of the goods receipt when receiving of nonconforming parts or materials from suppliers,
- as part of intermediate inspections or during the final inspection in current production,
- after delivery through returns (guarantees) or repair requests.

It does not matter whether the nonconforming product or service is still in the operational value chain or whether it has already been delivered. There is a need for action in both cases. In addition to limiting the damage, the focus is on troubleshooting through replacement or corrective (and service) measures to keep the damage to the customer as low as possible. 8D reports or the 5 W procedure are helpful for a systematic root-cause analysis and a determination of corrective actions. These tools are part of the standard repertoire of quality management.

Since nonconformities can occur anywhere and at any time in the value chain, it is even more important to have a structured approach in dealing with them, i.e. a control and monitoring process. This is the only way to minimise the risk of unintentional installation or delivery and recurrence.

A written documented procedure must be maintained to control nonconforming products and services. At this point many companies forget to define the approval of personnel entitled to make decisions on nonconformities. However, it must be defined how employees who decide on how nonconforming products are handled are qualified along with the conditions and the responsibility for authorisation.

Adequate records shall be kept of nonconforming products or services and associated measures, including special releases by one's own personnel, customers or national aviation authorities. Records must be kept regarding the type and extent of nonconformity, the decision-makers, performance changes or concessions as well as the (special) release.

Process

If a nonconform product or service has been identified, emergency measures must be taken first. This includes the segregation[38] and unambiguous identification of the objects concerned. Further activities can be a delivery stop, a work stop at the affected work step, or the blocking of the corresponding material batch.

Further troubleshooting measures must be taken (see Subsect. 8.7 a). This requires both a root cause analysis as well as the determination of the scope of the nonconformity (see also Subsect. 10.2). The organisation shall verify if the nonconformity is a one-off or a systematic occurrence. For this purpose, it must be checked if other products, services or processes are affected by the nonconformity. At the same time, measures for extended error containment must also be defined (cf. Subsect. 8.7 b and c).

If affected products have already been delivered, the customer must be informed and, in the worst case, a recall initiated in order to minimise the risk of damage. In this case, the addressee group must be determined in accordance with the second NOTE. In addition to the customer other relevant interested parties may also be considered (e.g. end customers, authorities, suppliers).

The assessment of the detected nonconformity and the decision on how to proceed must be made by qualified and authorised personnel. Minor nonconformities are often evaluated in practice by the certifying staff alone.[39] More significant errors, however, are usually investigated by an in-house team of experts (Material Review Board—MRB) of certifying staff/quality assurance, design engineering and product management, possibly including QMR as well as work or material planning departments. In addition, the customer must be involved in the decision-making process if the product no longer complies with the agreed specification after special release.[40] The latter must then agree to the decision.

Solutions for Handling Nonconforming Products or Services

The options for dealing with nonconforming products or services are:[41]

- correction or rework
- return to the supplier or scrapping.
- use-as-is or reclassification (e.g. due to restricted use) incl. special customer release or authority.

---

[38]If the problem cannot be solved immediately, the nonconforming product must be stored in the quarantine warehouse.

[39]If necessary, it should be noted that the certifying staff has been authorised by the design department to perform this activity.

[40]This should be documented with the customer's signature in order to avoid any recourse claims at a later date.

[41]With EN 9120, the handling of nonconforming products is limited to the following measures: 1) scrapping, 2) rejection to the supplier, 3) rejection to manufacture for re-validation, 4) submission to the customer or competent EASA Part 21 J operation for release for use in the as-is condition.

The following specific measures may also be considered:

- termination of the agreed service provision or only a temporary stop,
- replacement, renewed provision of services,
- offering or providing an alternative solution.

When opting for measures 1–3, the product or service is released by a qualified employee, e.g. the product-responsible design engineer or manager, by means of a special release. When opting for a correction or rework, the product or service must be re-verified prior to special release to ensure that it meets the requirements.

In the event of design deficiency or the use of nonconforming products in their actual state, the person responsible for design has to make the release decision. It may even be necessary to involve the responsible EASA Part 21 J design organisation in the release process in order to rule out violations of airworthiness regulations.

This applies analogously to production errors if the company is delivering to a production organisation in accordance with EASA Part 21G, here the customer can inform its National Aviation Authority, if necessary. Usually, this procedure is also fixed in contracts or in the GTC. The reporting to the customer should usually include the serial number concerned, a description of the nonconformity, possible effects and the procedure for correction.

If the product no longer corresponds to the specification after special release, the customer must be involved in the decision and must agree.

A product correction is not permitted in accordance with aviation legislation if:

- the products have a defect that cannot be repaired,
- parts do not correspond to the approved specification (Approved Data) and it is foreseeable that a conformity cannot be established,
- parts have a life limitation (life limited parts) and have exceeded their service life or complete documentation cannot be provided,
- an unauthorised, irreversible change has been made to the product,
- parts and materials were exposed to extreme conditions (e.g. thermal) and can no longer be returned to their original condition.

In these cases, the product can only be scrapped. Parts intended for scrapping must be identified conspicuously and permanently as such, marked unusable by destruction and disposed in an appropriate (special) waste container. The scrapping must therefore take place in such a way that the product becomes irretrievably unusable and remains easily identifiable as such so that it is excluded from further use (cf. Fig. 8.10). It is permissible to leave the scrapping and disposal to a qualified external specialist company.

As part of the certification audit, it must then be expected that the certification auditor wants to see the contract with the (external) scrap disposal company in order to ensure that the disposal is in line with the agreed requirements.

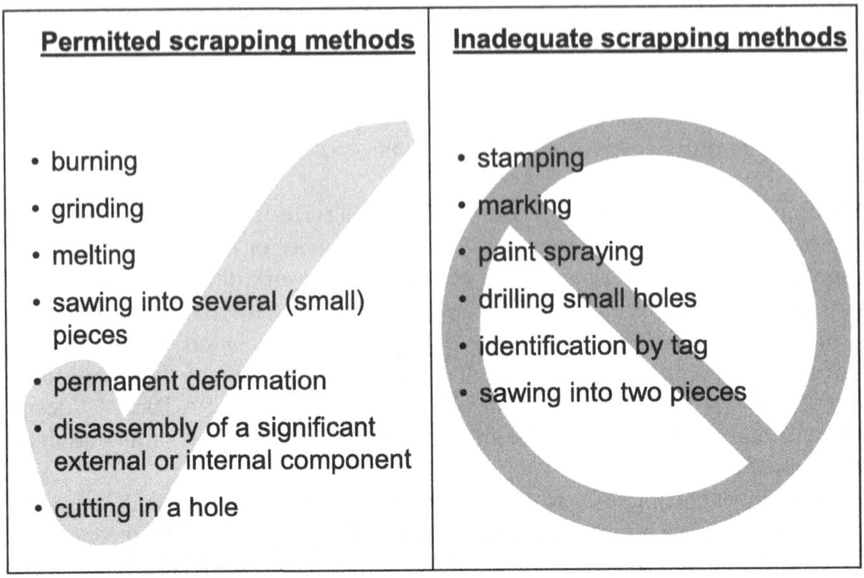

Permitted scrapping methods	Inadequate scrapping methods
• burning • grinding • melting • sawing into several (small) pieces • permanent deformation • disassembly of a significant external or internal component • cutting in a hole	• stamping • marking • paint spraying • drilling small holes • identification by tag • sawing into two pieces

Fig. 8.10 Permissible and impermissible methods of scrapping (cf. Hinsch (2019), p. 247)

If a counterfeit or suspect counterfeit part has been discovered rather than a nonconforming one, the part must be subject to a strictly controlled treatment and supervision. This is to minimise the risk of being placed on the market again.

Examples of Handling Nonconforming Products
Incorrect deliveries at goods receipt: Any defects detected during the goods receipt inspection are documented including a description of the non-conformity. The product receives a red tag or red tape blocking it from further use. If there is no immediate complaint, the goods will be stored in the quarantine area.

Carves/chippings from production: If the work is not carried out properly but the incurred damage has no effect on the functional characteristics, the certifying staff usually decides on the further procedure. If a technical change becomes necessary for the revision/repair, the product-responsible design engineer must be consulted for a decision.

Systematic errors in the companies' own production lead to a production stop. The production manager and the QMR must be notified. When searching for the cause and scope of the error, it must be checked if other products, systems and/or processes are also affected. If this is the case, these must also be identified, labelled and treated accordingly. The design engineer responsible for the product is usually involved in the evaluations and decisions. Corrective action shall be taken in accordance with Subsect. 10.2 (corrective action). Systemic errors can also be identified in the areas of goods receipt and repair e.g. due to incorrect order details or frequent errors in repair devices.

Errors in delivered products should at least be clarified in cooperation with the design engineer responsible for the product, the QMR and the account manager. If affected, further operational areas must be included. The nature and extent of the defect is to be determined. Subsequently appropriate measures are to be taken to rectify the nonconformity depending on its severity in conjunction with the customer (e.g. on-site repair, replacement of products or repair during the next planned service visit).

References

CEN-CENELEC Management Centre: ISO 10007:2003 Quality management systems. Guideline for configuration management. ISO 10007:2004-12 (2004)

CEN-CENELEC Management Centre: EN 9100:2018—Quality management systems—Requirements for aviation, space and defence organisations. Brussels (2018)

CEN-CENELEC Management Centre: EN 9120:2018—Quality Management Systems—Requirements for aviation, space and defence distributors. Brussels (2018)

CEN-CENELEC Management Centre: EN 9101:2018—Quality management systems—Audit requirements for aviation, space and defence organisations. Brussels (2018)

European Aviation Safety Agency—EASA: Acceptable Means of Compliance and Guidance Material to Part 21. Decision of the Executive Director of the Agency NO. 2003/1/RM (2003)

European Commission: Commission Regulation (EC) on the Continuing Airworthiness of Aircraft and Aeronautical Products, Parts and Appliances, and on the Approval of Organisations and Personnel Involved in These Tasks [Implementing Rule Continuing Airworthiness]. No. 2042/2003 (2003)

Federal Aviation Authority: FAA Unapproved Parts Notifications (UPN). In: http://www.faa.gov/aircraft/safety/programs/sups/upn/. Accessed on 31.01.2020 (2020)

Hinsch, M.: Industrial Aviation Management: A Primer in European Design, Production and Maintenance Organisations. Heidelberg/Berlin (2019)

International Aerospace Quality Group: IAQG 9100:2016 Clarification (Resolution Log). Rev. Date Oct. 2019 (2019)

International Organisation for Standardisation and International Accreditation Forum: ISO 9001 Auditing Practice Group: Guidance on: Design and Development Process (2016)

International Organisation for Standardisation and International Accreditation Forum: ISO 9001 Auditing Practice Group—Guidance on: Customer Communication (2016)

Performance Evaluation

9

Chapter 9 is dedicated to the control, analysis and evaluation of the service provision. This is intended to ensure that customer satisfaction as well as product and service conformity are maintained. The main instruments for evaluating performance are measurements and analysis, auditing and management evaluation.

9.1 Monitoring, Measurement, Analysis and Evaluation

9.1.1 General

In order to evaluate if the planned results have been met, the effectiveness of the value chain must be measured and monitored. The basis for this is the process performance. The measurements either confirm that the quality and performance objectives have been met or reveal weaknesses and deviations. There is a close resemblance to the standard, Sect. 9.1.3 *Analysis and evaluation*, which supplements the current section.

The nature, scope and frequency of monitoring and measurement must be defined and aligned with the size of the organisation and the service portfolio. Small businesses with simple value creation may only require a dozen measurements along with a few KPIs, while companies with group structures must have a comprehensive control and monitoring system. With regard to frequency, some measurements are to be carried out on an ongoing basis during the provision of the service (e.g. acceptance tests), while others only need to be conducted once every 3 months. The monitoring and measurement shall be carried out in accordance with

- the value of the processes for the service provision and
- the significance of individual product and service components (including those supplied) for the overall result.

© Springer-Verlag GmbH Germany, part of Springer Nature 2020
M. Hinsch, *Guideline for EN 9100:2018*,
https://doi.org/10.1007/978-3-662-61367-2_9

Most certification auditors do not accept an annual KPI measurement of the core processes as this long period does not allow weak KPI performances to be identified. Thus, those KPIs cannot fulfil their function as a fact-based decision-making aid. It is also important that measurements show a clear root-cause-effect relation. For this purpose, suitable measurement methods or key figures must be available and survey frequencies must be determined. Records shall be kept for monitoring, measurement and analysis. Detailed information on which basis is to be measured can be found in the list in Sect. 9.1.3 (a–f) of this book.

9.1.2 Customer Satisfaction

Quality is when the customer returns, not the product. EN 9100 therefore strongly emphasises on consistent customer orientation in addition to a strict process focus. Customer satisfaction is not limited to the product or the service but refers to all aspects and components of a business relationship. For a successful long-term market presence, companies must therefore ensure not only a competitive product and service portfolio, but also an appropriate delivery performance. The objective is finally, that customers are satisfied with performance *and* cooperation. From a standards perspective, it is irrelevant whether the customer is external or internal to the company. The latter are particularly common in large companies.

Customer satisfaction is regarded as an essential parameter for evaluating the performance of the QM system. Customer satisfaction must therefore be measured and monitored on a regular basis. EN 9100 therefore demands at least the following KPIs:

- On-Time-Delivery (OTD),
- Product conformity,
- On-target-quality, OTQ (e.g. through final acceptance tests or complaint rate),

These key figures are verified in every certification audit. This is not negotiable! It should be noted that customer satisfaction is audited in the certification audit as a process and not as a single requirement.[1] Thus, not only key figures are crucial, but also the procedure for related data collection and evaluation. Thereby, the output of this process is also an important input for other QM processes, e.g. for improvement, internal auditing or management review.

In business practice, customer satisfaction measurements all too often not given the attention they deserve and require. In many certified companies the voice of the customer is not systematically captured. It is then difficult to correct undesirable developments with structured measures. Although most CEOs and sales departments have a rough idea of their level of customer satisfaction, their

[1]cf. ISO 9001 (2016), p. 2.

opinion is often only based on estimations which is not an acceptable parameter for managers.

If potential for improvement in customer satisfaction is identified, systematic action must be initiated. It is therefore necessary to ensure that measures are derived and subjected to effective tracking. The results of these activities must be reflected in the management review.

As a note: Sending questionnaires to determine customer satisfaction is no longer state-of-the-art in many cases as the ISO EN standards have become widespread over the past 10-15 years. Since then, quality-related questionnaires have been used increasingly and are rarely answered thoroughly by respondents. Thus, these questionnaires increasingly irritate customers.

The NOTE in Sect. 9.1.2 of the standard lists alternative methods or parameters for measuring customer satisfaction. Furthermore, changes in business volumes, sales reports and, if necessary, evaluations of the type, scope and development of repairs are also suitable for evaluation. It is also possible to measure customer satisfaction based on returned goods, credit notes, claims under guarantees, warranty costs and invoice corrections.

Records shall be kept of the measurement, analysis and actions that improved customer satisfaction in accordance with Sect. 9.1.3. The retention period should be at least 5 years.

9.1.3 Analysis and Evaluation

This section aims to ensure that collected QM data are continuously analysed and evaluated. Structured data evaluation is used to derive direct assessments regarding the status of the product, service and process quality, customer satisfaction and finally the performance of the QM system in its entirety. The focus is on an analysis that enables fact-based decision-making.

All data which can provide information about quality, may be used as a source of information. In accordance with Sect. 9.1.3 of the EN 9100, data on the following topics must be evaluated:

(a) *Products and services*: e.g. result of product inspections, warranty costs, requests for corrective actions,
(b) *Customer satisfaction*: e.g. sales figures, type and number of corrective actions and customer complaints, surveys and customer feedback,
(c) *Effectiveness of the QM system*: e.g. period to close audit findings, cost of non-quality
(d) *Planning quality*: e.g. adherence to schedules or resource utilisation: plan to actual hours, on-time delivery,
(e) *Risks and opportunities*: e.g. plan deviations in hours/days, rework, downtimes,
(f) *Supplier performance*: e.g. determination of on-time delivery, incoming goods findings, costs, innovation capability,
(g) *Improvement*: e.g. past developments to the examples mentioned in a)–f).

Section 9.1.3 does not stand alone but must always be seen in conjunction with other sections of the standard. Those other chapters therefore provide a basis for the analysis and evaluations which required here: Examples of such inputs are in particular Sect. 8.4.1.1 (Supplier Performance), Subsect. 8.6 (Release of Products and Services), Sect. 9.1.1 (General Information on Monitoring and Measurement) and Sect. 9.1.2 (Customer Satisfaction).

The evaluated results either provide evidence of compliance with all quality requirements or form the basis for corrective and improvement measures (Sects. 10.2 and 10.3 respectively) and eliminating nonconformities (Subsect. 8.7). Moreover, the results of the data analysis are an important input for the management review (Subsect. 9.3).

9.2   Internal Audit

Companies with an EN 9100 certification must evaluate their processes by means of internal audits. This instrument serves the purpose of checking whether the processes are implemented in daily practice and meet the requirements of EN 9100 as well as all other requirements. The audit provides management with an instrument for a structured and independent investigation that provides information on the effectiveness and efficiency of the processes and the QM system.

The starting point for all audit activities is the audit program prescribed by the standard. It defines the foundation of the audit in terms of

- where (in which areas),
- at what intervals and to what extent, and
- with which type of (system, process or product audit)

audit must be conducted. The audit program thus fulfils the purpose of structuring internal auditing. It ensures that all components of the QM system are checked regularly. The audit program is a comparatively static document that is reviewed annually but is usually not or only slightly adjusted. The audit program is usually planned over a 3-year audit cycle.

Besides to the audit program, there is the annual audit plan in which dates, auditors, processes and departments are planned for auditing in the current year. In small and medium-sized companies, audit plans and audit programs often form a single document. The annual audit plan is a document with a forecast of 12 month.

The frequency of audits is often under discussion. The standard does not provide any concrete information on this. However, there is agreement among certification auditors that each chapter of the standard and each process must be audited at least once every three years. Core processes and quality management must be audited annually.

In order to increase the effectiveness of the internal auditing, it is recommended to define a KPI for this process. Since the number of audit findings says little about the process quality and as the auditor qualification can only be quantified to a limited extent, the measurement of the processing time for closing audit finding is recommended as a performance indicator. This should be less than 30–45 days.

Internal audits conforming to the norm must essentially be system audits, as this is the only way to assess the effectiveness of the QM system. However, it makes sense to integrate elements of a process audit with a focus on an order or products, as to reduce the degree of abstraction. This also increases comprehensibility and acceptance among the audited persons.

Finally, each audit is planned individually. The planning is subdivided into the following main elements:

1. audit preparation,
2. audit execution, and
3. audit follow-up and tracking of nonconformities (findings).

Audit Preparation

The first step is the preparation of the audit. This includes the necessary coordination with the process owners or the head of department, the definition of focal points and the announcement of potential changes in the schedule in order minimise the disruptions of day-to-day operations or important orders. Usually, an audit plan must also be created and distributed. During preparation, care must be taken to ensure that the results of previous audits are considered in order to intensify the audit focus in the context of previous findings.

Audit Execution

EN 9100 does not impose specifications regarding the audit method. The NOTE refers to ISO 19011, the standard for auditing QM systems. Alternatively documents from audit courses or textbooks, which usually offer equivalent, often more practice-oriented information may be used.

It is important that auditors are qualified to perform their duties. The quality of results and thus the effectiveness of audits mainly depends on the auditor's qualification. Therefore, it is to be expected that the qualification of the internal auditors is checked during the first certification audit, especially in SMEs. As the auditor often performs his or her task in personal union with the QMRs, this check also serves to get to know each other. The qualification should be demonstrated by a certificate from an internal auditor course and an EN 9100 training or by sufficient professional experience (although neither allow a direct statement to be made regarding the actual qualification).

As an alternative to employing auditors, it may be more feasible to resort to external auditors for 2–3 days a year due to cost and know-how. Ideally, the auditors are EN 9100 authorised auditors with sound aerospace audit competences. These are the most experienced aviation auditors on the market for this task. This is an advantage especially for SMEs because external auditors often have a different, more experienced view. Above all, their independence and impartiality are to be considered when using external auditors. This particularly applies to the area of quality management, as the auditor is not allowed to audit his or her own activity him or herself (Sect. 9.2.2 c). If QMR and auditor are the same person, a second qualified employee, who is not assigned to QM, must audit quality management processes.

If sites or facilities are excluded from the scope of the certificate, the standard does not require for them to be covered by the audit program and therefore do not have to be audited. The same normally applies to accounting and financial processes. Irrespectively, the management may have an interest in regular internal auditing that is detached from the EN 9100.

When carrying out audits, process KPIs should also be evaluated in accordance with NOTE. KPIs allow conclusions regarding the effectiveness of the QM system to be drawn. They also strengthen the awareness for process orientation.

Audit Follow-Up and Tracking of Complaints
Following the audit execution, the auditor shall prepare the audit report. Ideally, a consistent format should be used for this purpose. The following information should be included:

- General information (date, audited department, auditor, participants, audit basis),
- Summary of the audit/audit contents,
- Nonconformities (findings), potential for improvement, strengths if applicable,
- Signature of the auditor,
- If desired, signature of the management.

If not already done in the audit, corrective actions with deadlines and responsibilities are to be defined for the audit findings. The responsibility for the root cause analysis as well as for the preparation and implementation of corrective actions usually lies with the affected departments or process owner, not with the auditor. Companies that do not comply with this unwritten procedure must expect auditors to report fewer audit findings. Instead, actual nonconformities are merely expressed as potential for improvement or verbal warnings.

Corrections and corrective action must be implemented immediately, normally within two to four weeks. The auditor is responsible for monitoring the timely and effective closing of audit findings. The auditor must also act if major findings affect other processes and departments or may occur in a similar manner (cf. also standards Sect. 10.2.1 b 3). Audit findings are sometimes only implemented long time after the deadline has passed. If even warnings of the auditor have no effect, the auditor must escalate the incident to the management. If this does not work either, the QMR may inform the certification auditor during the audit. This can lead to an audit finding with regards to the effectiveness of the audit process. By doing so, the certification auditor can may act on behalf of the QMR.

The audit results shall be communicated to management individually or in a group. A minimum of an annual report is required for management review. In most small and medium-sized enterprises, however, the audit reports are also sent to the management immediately after they have been prepared. In addition, the reports are often signed by the management to prove that they have been taken note of.

To ensure a stable audit process, a documented procedure for the audit process must be defined.

9.3   Management Review

The management must regularly carry out so-called management reviews. This term is sometimes misunderstood, because it is not the management that is evaluated, but the management is supposed to assess the effectiveness of the QM system. The management review is intended to give the organisational management the opportunity to obtain an up-to-date overview of the status of operational quality management. At the same time, this review serves to instruct corrections and improvement measures in the QM system.

The standard does not define form and frequency of management reviews. They may be integrated in independent quality meetings or into other meetings (e.g. budget or strategy planning). Management reviews can be conducted monthly, quarterly or (at least) annually. Management reviews may also be set in individual intervals: For operational topics, monthly management reviews are carried out (status of objectives and KPIs) And for strategic Q-topics, additional half-yearly or annual management reviews are held (status of internal and external issues, risks and opportunities or audit overview). There is also no obligation to exclusively address quality topics in these reviews. The review can also have a freely chosen name. Particularly for SMEs, Q-oriented management reviews are held annually, usually about one to three months before the certification audit. However, the review can also be performed at the beginning of the year as a review of the past fiscal year. If the certification audit is not carried out immediately after the review the auditor has a comparatively easy means to check the degree of implementation of the measures instructed in the review.

The main task of the management review is to evaluate medium- and long-term QM-relevant topics and to deal with internal and external issues (see Subsect. 4.1 and 4.2 Context of the organisation).

In terms of content, the review reflects on the quality activities of the period under review (it is thus it is called a *"re-view"*). All QM-relevant measurement results are "put on the table" and critically reflected:

- How has the customer satisfaction developed?
- What was the development of the process performance (at least core processes)?
- To what extent has the quality of products or services changed since the last review?
- What lessons have been learnt from audits and have the findings been satisfactorily closed within a reasonable time?
- Have the quality objectives been achieved and if not, what measures have been taken?

The strict use of the Plan-Do-Check-Act cycle is important here. If objectives are not achieved, effective measures (not only alibi measures) must be derived and shown during the certification audit.

In addition, there is an obligation to reflect and evaluate operational risks. This can be done, for example, with the help of a risk matrix as shown in Fig. 6.1. If actions are required, deadlines and responsibilities for measures must be set and documented. In addition to risks, opportunities should also be analysed in a structured way. The focus is not only on market opportunities, but also on operational improvements. The evaluation can be carried out analogously to the risk matrix by representing opportunities.

The review also asses the evaluation of suppliers. Here, OTD, product conformity, special dependencies, quality deficiencies and other risk are to be addressed.

A management review is predominantly a review of the past observation period. Measures must be derived from the conclusions gained. Therefore, a management review must lead to an output. The minimum requirements are listed in Sect. 9.3.3. Accordingly, improvement measures and possible decisions with regards to the QM system and operational resources are to be instructed. If objectives or requirements have not been achieved or implemented, effective measures must be taken.

Although it is not explicitly prescribed by the standard, the management review also offers a good opportunity to review the quality policy for its applicability and to update it, if necessary.

About two to four hours are required for the annual management review. In addition to the upper management and the QMR, the second management level should also attend this meeting. Such a broad spectrum of participants underlines the management attention for quality and, in view of the frequency of this meeting, certainly does not overstretch operational capacities.

As presentation/documentation media, both MS PowerPoint and the MS Word formats are suitable. Regarding the structure of the management review, there is no mandatory sequence. However, it should be ensured that the requirements listed in the EN Sect. 9.3.2 can easily be found in the slides or the management review protocol.

It is important that the output is documented. The management review records are reviewed at each certification audit.

References

CEN-CENELEC Management Centre: EN 9100:2018—Quality management systems—Requirements for aviation, space and defence organisations. Brussels (2018)

CEN-CENELEC Management Centre: EN 9120:2018—Quality Management Systems—Requirements for aviation, space and defence distributors. Brussels (2018)

International Organisation for Standardisation and International Accreditation Forum: ISO 9001 Auditing Practice Group: Guidance on: Customer feedback. 13.01.2016 (2016)

Improvement

10

10.1 General

In accordance with the requirements of standards Subsect. 10.1, companies must, wherever possible, improve and continuously develop their products and services, as well as the QM system itself. This is intended to maintain and expand customer satisfaction and competitiveness. The topic of *improvement* should not be underestimated. Improvement is classified as so important by the EN 9100 that it has be explicitly anchored in the quality policy and quality objectives. This is not negotiable!

The enumeration and the NOTE of Subsect. 10.1 clearly formulate objectives and starting points for improvement. In addition to those improvements that target products and services from a technical background, the "typical" QM measures that are identified and controlled by the QMR and initiated by the management review are also considered as improvements.

However, good improvement measures which are suitable for certification do not necessarily have to be carried out under the control of the QMR. Thus, many valuable improvements take place on the operational level or are derived from observations of the everyday business life without the involvement of the quality department. Reorganisations, investments in personnel strength or qualification as well as infrastructure measures are also regarded as (strategic) improvement measures. Further examples are the reorganisation of production processes, the instruction of a training program or the decision to purchase a new, more powerful machine. A large number of different activities and measures are therefore suitable to meet the requirements of Subsect. 10.1.

Whether the continuous improvement takes place in a formalized manner or is largely based on verbal agreements, has a strong QM orientation or takes place

© Springer-Verlag GmbH Germany, part of Springer Nature 2020
M. Hinsch, *Guideline for EN 9100:2018*,
https://doi.org/10.1007/978-3-662-61367-2_10

informally under a different "name" is irrelevant to the standard.[1] Only the effectiveness of the measures and the improvement procedure themselves counts. In order to meet the standard requirements, a culture of willingness to change must be established. Meanwhile examples of successfully implemented improvement measures by taking small steps may be presented. No revolutionary improvements are expected. The concept of continuous improvement is based on small steps.

In certification audits the biggest problem is often that it is difficult for the management or the QMR to identify activities of the past year as improvement measures and to classify them as such. Many small measures often go unmentioned due to their self-evident nature towards improvement. Investments in new equipment, reorganisations or the training of employees are also not mentioned, although these are mostly improvement measures. However, the spectrum is broad and so it is better to name too many measures than too few. Finally, the certification auditor can interrupt or slow the auditee down at any time.

The right level of improvement is difficult for the auditor to quantify anyway. From the auditor's point of view, the improvement process coheres with the standard if the company identifies improvement potential and implements improvements according to its own planning, using the PDCA cycle.

10.2 Nonconformity and Corrective Action

In day-to-day operations,errors or incidents are quickly rectified in order to resume to normal production and meet performance targets as soon as possible. However, the underlying causes, for example repetitive patterns, correlations or similarities of errors are drifting out of focus.

An essential instrument of the improvement system thus loses its effectiveness, because in such an environment it is difficult to ensure sustainability or to make use of learning effects. Therefore, the standard requires a *systematic* and *sustainable* approach to eliminate nonconformities.

Since the intervention only takes place after a deficiency, this is a reactive procedure. Figure 10.1 shows an exemplary process illustration.

After identifying an error, a nonconformity or a customer complaint,[2] measures to confine the deficiency as well as for rectification and, if necessary, for prevention must be taken. The steps required for this are described in detail in Sect. 10.2.1 b):

1. If necessary, emergency measures must be taken.
2. The deficiency must be evaluated. Affected products and services are to be identified (e.g. batch or serial numbers, processed document or affected

---

[1]However, a formalised approach should become more important as the size of the organisation increases.

[2]In the following briefly: Error.

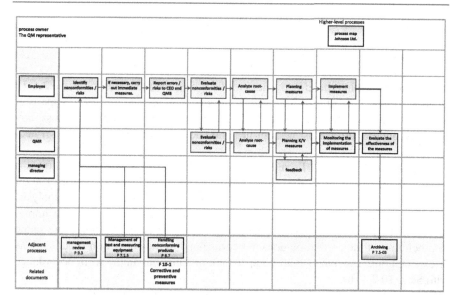

Fig. 10.1 Process flowchart corrective action

customer/order). The severity of the error must then be determined (e.g. extent or impact on customers).

3. It must be checked if other products or services do not meet the requirements due to the identified error (see also Subsect. 8.7).

4. The root cause of the error must then be determined. Possible are causes are for example production or design deficiencies, inadequate resource quality or quantity, errors in the specifications, human factors and many more. Only if the exact root cause of the error is known suitable measures can be taken to ensure appropriate correction and avoid recurrence. For an effective root cause analysis, recognised methods such as the Ishikawa analysis or the 5 W method are to be used. In this context, it must be checked whether the deviation has already had an effect or may have a potential effect on another product, process, employee, machine, etc. It must therefore be established whether the error is at random or if a systematic error has been identified. Error processing and documentation should be carried out using an 8D or 5D form. Therefore, it is to be determined if it is a random nonconformity or a systematic error. Error processing and documentation should be done using an 8D or 5D form.

5. After the root cause has been determined, measures for complete error correction and elimination follow (cf. Sect. 10.2.1 c). These measures may be for example:
   - Adaptation of processes,
   - Adjustment of the QM system,
   - Change of training content,

- Change of material specifications,
- Design changes,
- Change of supplier.

In order to determine the effectiveness of the corrective measures (Sect. 10.2.1 d), an effectiveness control must be carried out after implementation (e.g. comparison of data before and after the correction). During the implementation process, any risks and opportunities must be kept in mind and further measures derived if necessary (Sect. 10.2.1 e).

In contrast to ISO 9001, Sect. 10.2.1 g) of EN 9100 additionally requires a mandatory involvement of the supplier if he is responsible for the error or audit finding. A request for correction (including error description, root cause analysis, deadline) must be sent to the supplier. The supplier must then report the implementation by means of a so-called Corrective Action Report (CAR). If the scope or timeframe of the correction does not correspond to the correction specifications or expectations, assistance shall be provided, or sanctions imposed in accordance with Sect. 10.2.1 (h).

Too often systematic root cause analysis and the determination of sustainable measures do not take place in operational practice. The focus is usually only on the elimination of the symptoms, but not the actual root-causes. Companies that do not yet have the necessary know-how will have to place a stronger emphasis on the development and application of root cause analysis in the future. Systematic and in-depth root cause analyses are becoming increasingly important in certification audits and are more and more expected from EN-certified companies.

Human factors must also be explicitly considered in the root cause analysis as well as in corrective actions (e.g. by means of a separate field in the 8D report). However, not the employee responsible for the deviation but the human *factor* is of interest here. According to the standard it does not suffice if the employee responsible is only instructed or trained as a countermeasure. Finally, it should be ensured that the error does not occur again in the future caused by the individual employee that caused the error (human being) as well as with all employees (human factor). In this respect, the determination of the cause of human error usually requires thorough research and complex measures.

Finally, it must be ensured that the nonconformity itself, the measures taken, and their results are documented.[3]

There is an obligation to maintain a documented procedure for dealing with errors and nonconformities as well as their corrections. A process description must therefore be available.

---

[3]It is therefore recommended to keep a form for the processing and recording of corrective actions.

10.3   Continuous Improvement

This section of the standard is about systematically and actively improving the performance of the QM system and thus all processes involved in the value chain.

In order to identify improvements, audit results, management reviews and process measurements as well as other quality parameters may be consulted.

Where appropriate, QM tools such as benchmarks, lessons learned and FMEA analyses should be used to analyse improvement potential and derive suitable measures. Whether the continuous improvement takes place in a formalized manner, is largely based on verbal agreement, has a strong QM orientation or takes place informally under a different name is irrelevant in regards to the standard.[4] Only the effectiveness of the procedure on the improvement of the QMS is relevant. However, the most important measures and their effectiveness control must be recorded.

References

CEN-CENELEC Management Centre: EN 9100:2018—Quality management systems—Requirements for aviation, space and defence organisations. Brussels (2018)
CEN-CENELEC Management Centre: EN 9120:2018—Quality Management Systems—Requirements for aviation, space and defence distributors. Brussels (2018)

---

[4]However, a formalized approach should become more important as the size of the company increases.

The Process of EN 9100 Certification

11

For those readers who are not familiar with the certification or would like to get more information on individual aspects of certification, this final chapter offers an insight into the fundamentals of the entire certification process.

11.1 Preparation for the Initial Certification

The certification audit embodies a final, crucial stage when seeking EN 9100 certification. This is preceded by a long decision-making process by the management in which the pros and cons of certification are evaluated. In this phase, particularly the QMR is ought to familiarize him or herself with the content of the desired standard. Initially, however, it is not crucial to know and understand each individual requirement of the standard. The emphasis is placed on raising an awareness of the goals, tasks and implications of EN 9100 as well as its fundamental expectation on the QM system.

The prime source for familiarisation with the EN 9100:2018 is the text of the standard itself. This book also provides an excellent basis. If no experience with ISO EN quality management systems exists yet, seminars may be booked.

The decision in favor of an EN certification is followed by an approximately three to twelve months long period for the operational implementation of the EN 9100 requirements. In this second step, it is necessary to study the aerospace requirements in detail in order to be able to determine operational gaps. In order to achieve a clear starting point, it is advisable to hire an experienced (ideally authenticated) EN 9100 auditor for a delta analysis.

Based on the results, the creation of a requirements list (in Excel) can be helpful. Here the requirements which are already fulfilled based on objective evidence (documents, records, etc.) are marked as "done". Where deficits have been identified, deadlines and responsibilities for implementation as well as further comments are documented based on the analysis results.

© Springer-Verlag GmbH Germany, part of Springer Nature 2020
M. Hinsch, *Guideline for EN 9100:2018*,
https://doi.org/10.1007/978-3-662-61367-2_11

Experience has shown that the greatest need for action lies in the quality-oriented self-perception and the introduction of an appropriate quality culture. Also, there are often gaps in the formulation of quality objectives, process measurement and the associated KPI tracking. Other deficiencies may be:

- *Process orientation:* One of the core objectives of EN 1900 and EN 1901 is for certified companies to establish and maintain stable, controlled processes, i.e. a clearly defined value chain. A prerequisite for this is a fundamental understanding of process-oriented corporate management. This includes not only a description of the operational processes using flow charts, but more importantly a KPI control of the processes as well as the consistent use of the PDCA cycle. The type and scope of the process description are based on the size of the organisation and the provision of services. A process map with the operational core and accompanying processes is a fundamental requirement. All top-level processes must be represented as well as defined. This includes the definition of target values and the allocation of sufficient data (cf. also Book Sect. 9.2.3).
- *Documentation:* The structure of the documentation is in the form of a hierarchical pyramid. The top level of the QM documentation contains the written quality policy and the quality objectives as a base of the operational QM system. In most companies these are supplemented by a QM manual, although such a manual is no longer required by the standard. Process descriptions or procedural instructions are provided at the second level. In addition to solid training, only the written word, diagrams and documented visualisations can create an appropriate process understanding. A written procedure is generally recommended for the following processes, irrespective of the size of the company:
  - Management/leadership (Chap. 5),
  - Risk management (Sect. 8.1.1/Subsect. 6.1),
  - Personnel competence and qualification (Subsect. 7.2),
  - Dealing with documents and records (Subsect. 7.5),
  - Sales (in particular preparation of offers and conclusion of contracts) (Subsect. 8.2),
  - Design (Sect. 8.3) and configuration management (Sect. 8.1.2),
  - Supplier selection, approval and monitoring (Sect. 8.4.1),
  - Purchasing (Sects. 8.4.1 and 8.4.3) and goods receipt (Sect. 8.4.2),
  - Outsourcing (Sect. 8.4.2),
  - Core process of production or service provision (Sect. 8.5.1),
  - Handling of nonconforming products and services (Sect. 8.7), and
  - Corrective actions (Sect. 10.2).

Descriptions of these processes help with structuring the value chain because work/process steps and responsibilities are defined and assigned. Even small companies with less than ten employees must describe the above-mentioned processes. After all, it is about creating system stability and independence from individual people as far as this is possible. The tendency is that the larger the company and the more complex the value chain, the more documentation is required.

Scope and Applicability
Before a certification audit can be executed, the scope of the QM system according to EN 9100 must be determined. It must therefore be defined which sites (or parts of companies) and which product portfolio is to be certified. This scope of application is defined as the audit scope and is usually determined prior to or during the stage 1 audit in close coordination with the certification body. During the stage 1 audit the scope of applicability is coordinated with the auditor.

If chapters of the standard cannot be applied due to intrinsic product specifications or due to the service portfolio, individual chapters may be declared not applicable (cf. Subsect. 4.3). A typical example is the exclusion from Subsect. 8.3 (design) if a company does not carry out its own design activities (e.g. build-to-print organisations). The scope of application including omitted standard chapters must be documented. This is usually done in the QM manual.

Completion of the Preparation Phase: Internal Audit and Management Review
To complete the preparation for the certification audit, an internal audit must be carried out. This should be done about one to two months before the audit, followed by a management review. The internal audit does not only ensure that the requirements of Subsect. 9.2 are fulfilled but also to check whether all EN requirements have been implemented. At the same time, the management review shortly before the stage 1 audit can help to sensitize management to all important aspects of the quality management system and the EN 9100 requirements.

External Support
During the preparation for certification, an external consultant can provide valuable support and at the same time accelerate the implementation process. This is particularly true for small and medium-sized enterprises that have little experience or capacity to maintain an aerospace QM system. Each company must decide individually, if a consultant is generally necessary, if he or she should accompany the entire preparation phase or if support should only be called for special occasions, e.g. for larger shortcomings. It is also an option request consultancy only at the beginning and the end of the preparatory phase for the evaluation of the company's certification maturity.

When a consultant is used, care should be taken to ensure that he is experienced in the field of aerospace explicitly, ideally, he or she has attained EN 9100 authentication. The support of ISO 9001 consultants instead of an EN 9100 expert has already led to some unpleasant surprises in certification audits. Comparing ISO 9001 to this industry standard, there are not only differences in content but also in the audited aspects that are emphasised. Also 9100 certification audits follow stricter rules than ISO 9001 audits.

11.2   Selection of an auditor and a certification body

While implementing the EN 9100 requirements, a certification auditor and a certification company, a so-called certification body, must be selected at an early stage (about 4–8 months before the planned audit date). The focus should be on selecting an auditor who values the company's trust and meets its requirements.[1] Once the auditor has been selected, he will name one or more certification bodies for which he works. Certification without a certification body is not possible.

If there are no contacts to an authenticated EN 9100 auditor or an experienced aerospace consultant, a certification body with an EN 9100 approval should be contacted. In Europe, the following companies dominate the market for certification according to EN 9100, although the market presence varies greatly between countries:

- AENOR
- AFNOR,
- BSI,
- Bureau Veritas,
- DEKRA,
- DNV GL,
- Lloyd's Register
- SGS, and
- Several German TÜVs.

At the beginning of the certification process there is often, but not always, a preliminary face-to-face meeting or a phone call with the auditor or sales manager of the certification body. Initial information relevant to the certification is exchanged and basic activities are defined, which ultimately serve the desired certification. In order to give the auditor or account manager of the certification body

---

[1]A consultant can also offer valuable help here, because he usually knows the certification auditors as well as the operational requirements.

an impression of the company, the following topics should be addressed at least briefly in this preliminary discussion:

- Business activities and products, including a short company tour if necessary/ possible,
- Type and scope of outlocated work/processes,
- Status of implementation and rough timetable for certification,
- the planned scope of the certification (sites or areas),
- planned non-applicabilities (exclusions, e.g. design or special processes).

In the preliminary discussion, the auditor or customer service representative should provide rough information on the certification process and the audit cycle (cf. Fig. 11.1), which consists of the following main stages:

- Stage 1 audit (pre-audit),
- Stage 2 audit (main audit),
- two smaller surveillance audits at annual intervals, and
- re-certification audit (after 3 years, extend similar to stage 1 audit).

In the selection procedure care should be taken to ensure that the certification auditor is located close to the region in order to keep travel costs low.[2] The location of the certification body, however, does not play a role.

The costs of the certification audit strictly depends on the size of the company (employees, number of sites) minus any exclusions (e.g. design). Only the daily rates and other costs differ slightly between certification bodies. For a company with 46–65 employees and a single location a total of 8000–10,000 EUR (6 audit days) including fees, can be expected for the initial audit. Distributors according to EN 9120 as well as companies excluding design can expect somewhat lower costs.

Proposals from the certification bodies always include the costs for a certification cycle of three years. Certification bodies therefore always offer four audits for an initial certification. The stage 1 audit (usually 1 day), the stage 2 audit (5 days, in the above example) and two less extensive surveillance audits in the following two years (2–5 days each).

11.3 Stage 1 Audit

For an initial certification, the main audit must be preceded by a stage 1 audit.[3] This is a mandatory preliminary audit to determine whether the company is prepared for the main certification audit. The general requirements for stage 1 audits

---

[2]For smaller companies, travel costs can easily exceed the actual audit costs.

[3]At this point only the most important aspects of the Stage 1 audit will be highlighted. Details can be found in Sect. 4.3.2 of EN 1901.

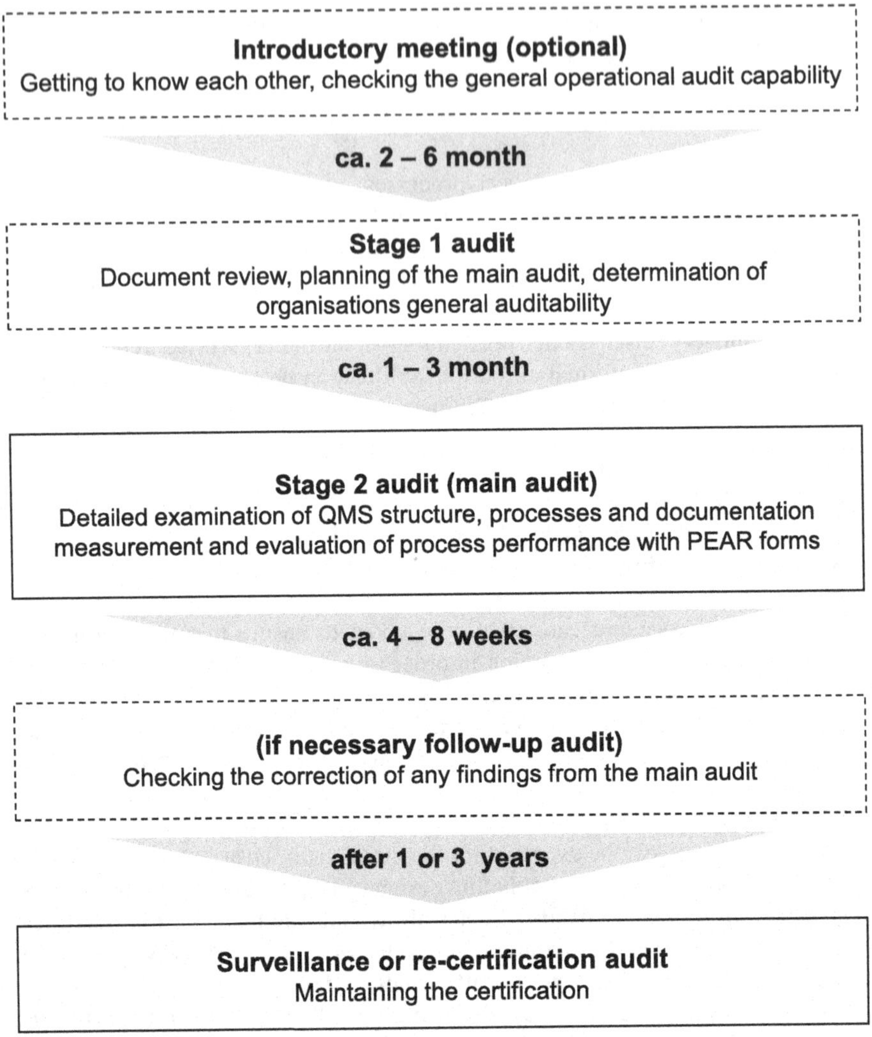

Introductory meeting (optional)
Getting to know each other, checking the general operational audit capability
---

ca. 2 – 6 month

Stage 1 audit
Document review, planning of the main audit, determination of
organisations general auditability
---

ca. 1 – 3 month

Stage 2 audit (main audit)
Detailed examination of QMS structure, processes and documentation
measurement and evaluation of process performance with PEAR forms

ca. 4 – 8 weeks

(if necessary follow-up audit)
Checking the correction of any findings from the main audit
---

after 1 or 3 years

Surveillance or re-certification audit
Maintaining the certification

Fig. 11.1 Process of a three-year audit cycle

are defined in ISO 17021 Sect. 9.1.10. The aerospace-specific aspects are defined in Sect. 4.3.2 of EN 1901.

In order to evaluate the company's quality and certification maturity, the auditor must first gain a general overview. An important activity of the stage 1 audit is a site inspection at an early stage. Such a tour is mandatory for this audit, as it is the best way to give the auditor a first impression of the premises and equipment as well as the production and working conditions.

The most time-consuming part of the stage 1 audit however is the evaluation of the documents. For this purpose, the auditor acquaints himself with the documents and objective evidence for each audit chapter. Particular attention is paid to quality objectives, KPI measurement, risk and process orientation, as well as to customer and legal requirements. The certification auditor gets a picture of the condition of the QM system by inspecting the following documents and records in particular:

- QM manual, process maps and process descriptions (usually requested by the auditor prior to the audit). The main objective is to identify core processes, process interactions and the outlocated processes.
- KPIs for process measurements as well as trend analyses with which the process effectiveness can be proven. At least data on product and service conformity, on-time delivery (OTD) and customer complaints of the past six to twelve months must be available.
- Customer satisfaction analyses (e.g. surveys, evaluations of customer visits, etc.).
- Process verifications, e.g.: supplier evaluations, risk FMEAs, training plans, measuring equipment lists, project plans, insight into the ERP system, QM to-do list, example 8D analysis, QAP.
- Records of the last management review.
- Internal audit reports of the last 12 months.
- In addition, the 3–5 largest customers in the aerospace and defence sectors incl. the percentage of sales (in percent!) as well as the number and proportion of employees are to be named.

The audit stage 1 report, which can be downloaded free of charge from the IAQG website (http://www.sae.org/iaqg/forms/index.htm), provides a comprehensive overview of the aspects to be addressed in stage 1.

The stage 1 audit also serves the purpose of planning the main audit and determining the audit program for the 3-year certification cycle. The certification auditor shall ensure that each standard EN 9100 requirement within the three-year certification cycle is audited at least once after the first year. For example, it can be specified that in the first surveillance audit only the production and service provision (Subsect. 8.5) are to be audited and that design (Subsect. 8.3) is not to be audited, while in the second surveillance audit production and service provision are not inspected while design is to be audited instead.

Finally, any non-applicable requirements (exclusions), including justification, must be determined no later than at stage 1 of the audit. The same applies for the scope of the certification. In addition, the auditor asks for the name of an OASIS database administrator[4]. If not done by then, an OASIS account is usually created together with the auditor.

---

[4]In the OASIS database, the certificates are administered online for all participating companies and made publicly available. However, once a database administrator has been appointed no further actions are required (except updating contact details and address, and at least one login per year). EN 9100 certificates from suppliers can easily be monitored via the OASIS database. It is therefore no longer necessary to ask the auditor to send you the valid EN 9100 certification. The auditor will explain in detail how to use the OASIS database.

As part of all these activities, the Stage 1 audit also offers the opportunity for the auditor and QMR or management to get to know each other.

During stage 1 of the audit, the certification auditor usually identifies some areas of concern such as insufficiently implemented standard chapters or missing process measurements. However, these are not formulated as audit findings in the Stage 1 audit. Such shortcomings must be eliminated before the main audit. Therefore, stage 1 audit should take place two to four weeks before the main audit.

11.4   Certification Audit (Stage 2)

Every certification audit, be it the initial audit, the surveillance audit or the recertification audit, is always structured and carried out according to a defined procedure. The core elements are:

- the opening meeting,
- the conduct of the audit,
- audit evaluation and the preparation of audit records, and
- the closing meeting.

Opening Meeting

The opening meeting is at the beginning of the onsite audit process.[5] Usually, the CEO gives the floor to the certification auditor immediately after the opening of the meeting. A round of introductions usually follows after which the auditor describes the certification assignment. In addition, he explains the audit procedure based on the audit plan, clarifies any ambiguities and asks for endorsement or any changes required. Smaller adjustments or a change in the sequence of the audit plan are usually no problem.

In addition, other legal and organisational aspects will be clarified, such as the availability of the room for the audit-team, whether the auditor will be accompanied and the necessity of safety equipment. At this point, lunch and other breaks are usually coordinated as well.

The CEO and the QMR as well as any other QM employees involved should be present at the opening meeting. It is advisable for the most important representatives of the second hierarchy level (e.g. head of production, purchasing and design) to participate in the opening meeting because their daily routine can be noticeably influenced for the duration of the certification audit.

As the CEO is at the head of the company and therefore occupies an important position for the certification audit, he is ought to set a positive example. This is because the appearance of the top management and their attitude towards the auditor during the certification audit, already provide information on the quality

---

[5]The requirements for the opening meeting are formulated in detail in Sect. 6.5.1 of ISO 19011.

maturity of the entire company. Therefore, the CEO should be alert and actively engaged in at least in the opening and closing meetings. However, it is not necessary for the CEO to constantly accompany the audit. Even for small companies this is not expected by the auditor.

The question of whether the auditor should be invited to dinner with the management when audit last several days, is sometimes a source of uncertainty. In a long-term audit relationship this is sometimes not a bad idea. Otherwise it remains the decision of the company. From an auditor's perspective, an invitation is rather not expected or desired, as the auditors also have to catch up with their day-to-day business and respond to emails after eight hours of concentrated work.

Audit Execution

The main objective of the audit is to check the compliance of the operational processes with the requirements of EN 9100, the customer, the legislator and the authorities as well as the requirements of any interested parties. Therefore, information is collected and evaluated by means of random sampling. This is done through interviews and observations as well as by reviewing records and documents (documented information).

Although the auditor himself decides on his audit procedure and audit focus, the EN 1901 specifies the minimum scope of an audit, regardless of whether it is an initial, surveillance or recertification audit. The elements that every audit must include among others are:[6]

- auditing of the processes according to the audit program,
- an examination and evaluation of the performance of the QM system, especially continuous improvement, changes to the QMS since the last audit, follow-up measures from previous audits,
- checking the status of customer requirements (changes, supplements, new),
- an evaluation of the process performance,
- audit of the procurement process, and
- a review of customer satisfaction trend and activities derived from it.

Every certification audit always includes an interview with the CEO. In this conversation, the CEO must be able to demonstrate his or her self-image in terms of quality and customer orientation. In addition, the CEO must provide information on the current status of the performance of the QM system and on measures for continuous improvement. Management should be able to report on the results of the latest management review. The discussion with the CEO should also show how and to what extent he is involved in the creation or updating of the quality policy and quality objectives. In order to obtain information on customer orientation and customer expectations, the CEO should also be able to explain his role in the process of customer satisfaction.

---

[6]See EN 1901 Sect. 4.2.2.1.

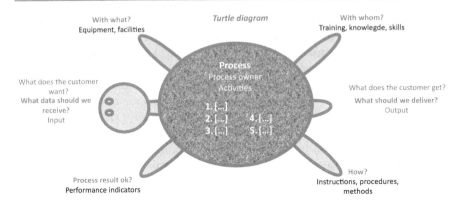

Fig. 11.2 TURTLE diagram

Another important element of any certification audit is the evaluation of the process orientation, including an assessment of process performance and effectiveness. In order to meet this requirement, audits are carried out in a process-oriented manner. For this purpose, the auditor takes a sample (e.g. a customer request or an order) at the beginning of the process and tracks it as far as possible over the entire process flow. During the audit, the auditor will pay special attention to whether:

- the process requirements have been met,
- all important elements of the process are described and applied,
- sufficiently documented information is available (in addition to the process description itself, work instructions, forms or checklists as well as records/ objective evidence),
- responsibilities and authorisations have been defined and are known to those affected,
- the process interactions/interfaces are sufficiently considered (in the documentation, as well as during work execution in the operational everyday life), and
- indicators are available to determine process performance and effectiveness.

A useful tool to determine the completeness of processes is the use of a Turtle Diagram as illustrated in Fig. 11.2. Furthermore, the diagram helps the certification auditor to fill in the PEAR forms.

In the end, it must be evident to the certification auditor that the processes meet the requirements and that they are effectively implemented and maintained (i.e. achievement of the desired results).

Auditors are also obliged to document their work and thus make it traceable. Each auditor shall record objective evidence for this purpose. For this reason, certification auditors take notes on an ongoing basis during the entire audit. Therefore, the auditor will permanently ask for order or project numbers, file and document names and other objective evidence with which the performance of the work can be confirmed. Among other things, the auditor will prepare evaluations of the achievement of process objectives in the so-called PEAR forms and the QMS matrix.[7] Constantly taking notes during the interview therefore has neither a positive nor a negative meaning.

At the end of each audit section, the auditor gives his thanks the auditees and gives a preliminary summary of the results. For certification audits lasting several days, it is common to conclude each audit day with a meeting. Participates of these meetings are usually the QMR, the CEO and the executives involved in the course of the day. In addition to an explanation of the daily results, such meetings provide an opportunity to discuss inconsistencies and any changes in audit plan of the next day.

Closing Meeting

Every audit end with a closing meeting[8]. The circle of participants usually corresponds to that of the opening meeting. In the final meeting, the auditor presents the audit results. This includes not only explaining any findings, but also mentioning positive aspects. If nonconformities have been identified, the auditor will explain the follow-up procedure and deadlines.

Following the explanation of the audit results, the auditor provides information on his certification recommendation. The auditor himself is not authorised to make the final decision on the outcome of the audit. The decision to issue a certificate is the sole responsibility of the certification body. The auditor only gives a recommendation. In practice, however, there is practically no difference between the auditor's recommendation and the final decision of the certification body.

In the final meeting, the auditor must also inform about how complaints about him or a co-auditor may be filed.

If this has not yet been done, the certificate's text of the scope is agreed in this final stage of the audit. In addition, other details for the certificate in terms of size, language(s) and the number of certificates to be issued must be clarify. For this purpose, the company should have prepared the certificate text in all desired languages in advance.

If no findings have been declared, the audit is usually concluded with the final meeting for the company. If findings have been identified, the auditor creates a nonconformity report (NCR) for each finding. This is done electronically via a workflow on OASIS, which is then processed online. Further details on the processing of audit findings are explained in the following subchapter.

Following the audit, the auditor must prepare the audit report and send it to the company. After this is done, the certification audit is completed for all participants.

---

[7]Caution: The related data may be confidential. Since the PEARs will be uploaded and archived in the OASIS database, their confidentiality is not guaranteed. The company must therefore decide whether actual KPI values may be documented in the PEAR. Alternatively, comparative KPI values (i.e. percentage changes) can also be used. Usually, the auditor can also be asked to document his evaluation without detailed information (e.g. "Process objective achieved," or "Objective not achieved", possibly including the name of the supporting documentation or file).

[8]The requirements for the final discussion are formulated in detail in Sect. 6.5.7 of ISO 19011.

Table 11.1 EN-Chapters/Processes with the highest audit findings

EN 9100 subchapter	Content
8.4	Procurement
8.5	Production and service provision
9.1	Monitoring, measurement, analysis and evaluation
7.5	Documentation requirements
4.4	QM system and its processes
8.1	Planning of product realisation
7.1.5	Monitoring resources
7.2 and 7.3	Competence and awareness
10	Improvement

11.5   Processing of Audit Findings

Companies must meet a wide range of requirements at all operational levels. These are imposed on them by EN 9100, by legislator or authorities, by customers or by interested parties. It is not always possible to implement all the required specifications in day-to-day operations as it is called for by the requirements. It is an essential element of a certification audit to identify such nonconformities. If a nonconformity (finding) has been identified it must be declared by the auditor. Findings are expressed in stage 2 audits as well as in surveillance and re-certification audits (cf. Table 11.1). In stage 1 audits no findings are declared. Instead areas of concern are being addressed. If they are not processed, they will be re-formulated to findings in the stage 2 audit.

Not all nonconformities have the same severity. According to Subsect. 6.8 of ISO 19011 there are the following classifications:

- major nonconformity,
- minor nonconformity, and
- recommendation/improvement.

A *major nonconformity* exists when the nonconformity of a requirement[9]

- leads to a failure of important elements of the QM system,
- there is a risk that a nonconforming product or service is processed or when there is a risk that a nonconforming product was or will be delivered or
- where the nonconformity is likely to have a significant effect on the customer by limiting the usability of the service or product.

---

[9]See EN 1901 Subsect. 3.3. and 3.4.

Findings are classified as major, for example, if KPIs have not been defined for the core/product realisation processes (=> systematic failure of an essential element of the QM system). Another example of a major finding is the absence of necessary inspection points or missing tolerances in goods receipt or in production. Here, the non-fulfilment of the requirement can have a direct influence on the customer when delivering a nonconforming product.

A minor nonconformity occurs when *a* requirement is not met but the severity does not justify a classification as major nonconformity. Findings are declared as minor in case of isolated errors or if the nonconformities of individual requirements are without substantial or lasting influence on the QM system, on the processes or on the product or service. Typical examples of minor findings are the omission of activities due to a lack of attention as well as procedures that have been executed incorrectly, intentionally or unintentionally. This includes the use of shortcuts and simplifications. If, however a minor finding occurs repeatedly across an organisation, the classification of the finding may be upgraded to a major nonconformity as this can lead to a failure of important parts of the QM system.

Findings are processed online via the OASIS database. For this purpose, the auditor creates a separate nonconformity report (NCR) for each finding during the certification audit. In OASIS, the auditor describes the finding including the non-fulfilled requirement and names the associated objective evidence. Furthermore, he classifies the finding as a major or minor nonconformity.

If the description of the finding is short, the company should either suggest a more detailed statement or make its own records. Often it is difficult for the responsible employees to reproduce the details or background information regarding a nonconformity a few weeks after the audit when the audit finding is being rectified.

In order to close the finding, the nonconformity must first be rectified at the location of the finding. The error must therefore be *corrected*.

In addition, the recurrence of the finding is to be prevented by a general, generic corrective action. The root-cause of the error must therefore be eliminated. It is the responsibility of the organisation to identify the root-cause of the nonconformity and to develop and implement corrective actions for the rectification of the finding. All too often the affected employees do not conduct a thorough a root-cause analysis, instead systematic errors are classified as minor one-off failures. The motivation for this is often convenience as this allows the finding to be closed quicker. However, the root-causes should be corrected, not the symptoms. Therefore, at least for major finding, a methodically sound procedure (e.g. 5 W method) must be demonstrated which verifies that the actual root-causes of the nonconformity have been identified. If the root-cause analysis is inadequate, it must be expected that the certification auditor will reject the rectification of the finding as inadequate. If he accepts an inadequate root cause analysis, the auditor himself must expect a finding or warning from his certification body.

An adequately analysed root-cause and the implemented corrective action must be documented online in OASIS and, if necessary, objective evidence must be uploaded. Rectification via NCR must be provided within 30–45 days after the end of the audit. An exception applies if the auditor demands immediate action due to potentially serious impact of the finding: in this case, the correction must be implemented within 7 calendar days.[10] However, it is not permitted to *close* the finding during the audit.

Once the organisation has sent the completed NCR(s) to the certification auditor, the auditor is responsible for evaluating the corrective action(s). Under certain circumstances, mostly when 6–8 or more audit findings have been identified, a follow-up on site audit is required. However, normally the correction of the findings can be evaluated based on the information on the OASIS NCR form sheet. To do so further documents are to be uploaded (e.g. action plan, updated process descriptions, records).

Only after all audit findings have been processed and reported back via NCR to the auditor, he may recommend the issuance or renewal of the certificate to his certification body. After this evaluation by the certification auditor, the entire audit documentation is cross-checked for correctness and completeness by a second EN 9100 auditor in a so-called veto audit. This takes place in the background and the company will not take any notice of this. For this reason, the certifier usually needs another two to three weeks after the completion of all on-site activities or after closing the last audit finding in order to issue the certificate. If this takes longer, the certification body should be contacted.

Note: If a comparable finding is identified again in the next year's follow-up audit two major nonconformities will be issued as this shows a lack of effectiveness of the corrective measures (cf. EN 1901 Sect. 4.2.2.5). One major finding is due to the renewed failure to meet the requirement, the other is due to the ineffectiveness of the correction process.

11.6   Surveillance and Re-Certification Audits

Surveillance Audit

The surveillance audit is conducted annually for two subsequent years after an initial or recertification audit. Here the QM system is not audited in full, so that the effort is about 50% lower than for the re-certification audits conducted every 3 years. The areas not audited in the first surveillance audit are then to be checked in the second surveillance audit. For example, it can be specified that in the first surveillance audit only the production and service provision (Subsect. 8.5) is to be investigated and that design (Subsect. 8.3) is not to be audited, while in the second surveillance audit production and service provision are not to be taken into account and design is to be audited instead. The scope of the surveillance audit is based on the audit program for the certification cycle defined in stage 1.

---

[10]See EN 1901, Sect. 4.2.4.

In each surveillance audit, the certification auditor must examine the core elements of the QM system.[11] This always includes the management review, the process effectiveness and performance, the On-Time Delivery (OTD) and customer satisfaction measurement. Additional areas of examination during the surveillance audit are the implementation of follow-up measures from the previous certification audit and changes to the QM system that have been made since the last certification audit. In addition, Subsect. 8.4 (Procurement) of the standard must be audited in each surveillance audit.

Re-certification Audit

The re-certification audit takes place every three years and roughly corresponds to the initial audit in terms of complexity (stage 2 audit).[12] The re-certification audit serves to renew the certificate. In contrast to the surveillance audit, the fulfilment of all standard requirements is checked in this audit. Furthermore, the focus of this audit is also on reviewing process effectiveness and performance, evaluating the ability to deliver conforming products and services, and evaluating customer satisfaction. In addition, the management review, the implementation of follow-up measures from the last certification audit and changes made to the QM system since then will be checked in detail.

Failing to demonstrate essential elements of the QM system or repeatedly failing to rectify audit findings results in a suspension of the certificate.

A recertification audit is followed by two annual surveillance audits.

The re-certification audit should be scheduled at least three months before the expiration date of the current certificate to avoid its expiration. The certification auditor ensure that the audit is carried out in time.

Audit for a Special Reason

In additionally to planned surveillance and re-certification audits, audits may be arranged for a particular cause. This may be due to a change of the certification body or an extension of the scope (e.g. new sites or the applicability of previously non-applicable EN chapters) outside the existing certification cycle. Usually, however, these events or changes may be addressed during the next regular certification audit.

On the rare occasion of the receipt of a severe customer complaint, an audit with a particular cause may be conducted. In this case, the company must undergo a separate audit and cover the costs if the certificate is not to be withdrawn.

---

[11]The requirements for the surveillance audits are defined in Sect. 4.3.4 of EN 1901 and are based on the requirements of ISO 17021, Subsect. 9.3.

[12]The requirements for the re-certification audit are formulated in Sect. 9.4 of ISO 17021; specifics for EN 9100 can be found in Sect. 4.3.5 of EN 1901.

References

CEN-CENELEC Management Centre: ISO 19011:2018—Guidelines for auditing management systems, Brussels (2018)

CEN-CENELEC Management Centre: ISO/IEC 17021-1:2019-12—Conformity assessment—Requirements for bodies providing audit and certification of management systems, Brussels (2016)

CEN-CENELEC Management Centre: EN 9101:2018—Quality management systems—Audit requirements for aviation, space and defence organisations. Brussels (2018)

CEN-CENELEC Management Centre: EN 9100:2018—Quality management systems—Requirements for aviation, space and defence organisations. Brussels (2018)

Index

© Springer-Verlag GmbH Germany, part of Springer Nature 2020
M. Hinsch, *Guideline for EN 9100:2018,*
https://doi.org/10.1007/978-3-662-61367-2

Printed in the United States
By Bookmasters